MATH 76

An Incremental Development

MATH 76

An Incremental Development

Stephen Hake

John Saxon

SAXON PUBLISHERS, INC.

Math 76: *An Incremental Development*
Second Edition
Teacher's Edition

Printed in the United States of America.

ISBN: 0-939798-75-1

Editor: Nancy Warren

Fifth printing: March 1996

┌─── *Reaching us via the Internet* ───┐

WWW: http://www.saxonpub.com

E-mail: helpdesk@saxonpub.com

Saxon Publishers, Inc.
1320 West Lindsey
Norman, Oklahoma   73069

To Mary, Bryon, Elizabeth, John, James, and Robert

Contents

Preface

To the Student

This book will take you from arithmetic to the threshold of the wide world of mathematics. The lessons encompass the content of arithmetic—addition, subtraction, multiplication, and division of whole numbers, decimal fractions, and common fractions. However, the lessons go beyond the topics of arithmetic to include foundational topics of geometry, measurement, algebra, number, and scale and graph reading. Word problems include items from history and science as well as from the realm of imagination. Working through the pages of this book you will find similar problems presented over and over again. **Solving these problems day after day is the secret to success. Work every problem in every practice set and in every problem set. Do not skip problems. With honest effort you will experience success and true learning which will stay with you and serve you well in the courses which follow.**

To the Teacher

This book grew out of a decade of intense classroom interaction with students in which the goal was for students to learn and **remember** the foundational skills of mathematics. The term "foundational" is appropriate because mathematics, perhaps more than any other subject, is a cognitive structure which builds upon prior learning. The ultimate height and stability of the mathematical structure within each individual is determined by the strength of the foundation. This book, as well as each book which precedes and follows it, provides the student with the time and opportunities necessary to build a rock-solid foundation in beginning mathematics. **For this to occur it is essential that all practice problems and all problem sets be completed by the student.**

How to Use This Book

This book presents a series of daily lessons, each followed by a set of 25 problems. Rather than providing practice in only the new topic, the problem sets review everything that has been taught previously. The sequence of the lessons and the content of the problem sets have been carefully planned. Do not skip lessons or sets. The teacher should briefly present the lesson, using the examples to lead the students through guided practice. The students should do the "Practice" at the end of each lesson as guided or independent practice before going on to the problem set. **It is essential for students to work every problem in the lessons and in the problem sets.** All work should be shown and all errors should be corrected. Consistent, honest effort will produce genuine learning with a high level of retention.

Many teachers have found the following suggested class schedule effective.

- Begin class at the bell with a "bell ringer" activity such as a speed test on basic facts or a short ungraded quiz (5 minutes).
- Present the new lesson and monitor the Practice section of the lesson. (Most lessons can be covered in 10 minutes or less.)
- Read or display answers to homework (3 minutes).
- Reserve the *majority* of class time for students to work on the new assignment and to correct errors quickly from the homework problem set. During this time the teacher should assume the role of a tutor. Designating student tutors or cooperative groups is also helpful. Students should be strongly encouraged to work on the problems that they find to be difficult while in class.

testing   An available test booklet contains two forms of tests for every five lessons. Tests should be given about five lessons after the last concept has been taught. Thus Test 1, which covers topics from Lesson 1 through Lesson 5, should be given after Lesson 10. Test 2 should be given after Lesson 15, Test 3 after Lesson 20, and so on. This allows the students time to learn the new topic before being tested on it. Students will make excellent progress if they are able to score 80% or better on the tests. Students who fall below the 80% level should be given remedial attention immediately.

The appendix contains practice problems grouped by specific topics. These problems are intended for remediation and can be skipped entirely if students are making satisfactory progress. They are made available for use, in whole or in part, as a supplement to, not as a replacement of, the regular problem sets.

Since all mathematical learning is built upon prior learning, success in this book should be considered a prerequisite for starting the book which follows. It is recommended that a student who consistently scores below 70% on tests be required to repeat the book the following year.

standardized tests Students who successfully complete this book usually score very high on standardized tests. However, tests are often given in spring before students have had the opportunity to complete the book. Some tests include measuring with a protractor and locating points on the coordinate plane. These two topics are covered in Lessons 139 and 140. These lessons are relatively independent and may be presented earlier in the year along with the specific drill on these topics from the appendix.

Stephen Hake　　　　　　　　　　　　　　　　*John Saxon*
El Monte, California　　　　　　　　　　*Norman, Oklahoma*

Adding Whole Numbers, Adding Decimal Numbers and Money

adding To combine two or more numbers is to **add**. The numbers which are added together are called **addends**. The answer is called the **sum**.

When adding, we should remember that we can only add like things. As the saying goes, "You can't add apples and oranges." When adding numbers we add digits which have the same place value.

example 1.1 $345 + 67 =$

solution When we add whole numbers, we write the numbers so that the last digits are aligned one above the other. Then we add.

$$
\begin{array}{r}
3\,4\,5 \quad \text{addend} \\
+\quad 6\,7 \quad \text{addend} \\
\hline
\mathbf{4\,1\,2} \quad \text{sum}
\end{array}
$$

example 1.2 Monica had $5.22 and earned $1.15 more. How much money did Monica have in all?

solution The words **in all** tell us that we should add. When we add decimal numbers, we write the numbers so that the decimal points are aligned one above the other. This makes us add digits whose place values are the same.

$$
\begin{array}{r}
\$\,5\,.\,2\,2 \quad \text{addend} \\
+\,\$\,1\,.\,1\,5 \quad \text{addend} \\
\hline
\mathbf{\$\,6\,.\,3\,7} \quad \text{sum}
\end{array}
$$

checking One way to check addition is to change the order of the addends. The order of the addends does not matter when we add.

$$
\begin{array}{r}
\$\,1\,.\,1\,5 \\
+\,\$\,5\,.\,2\,2 \\
\hline
\mathbf{\$\,6\,.\,3\,7} \quad \text{check}
\end{array}
$$

practice **a.** $3675 + 426 + 1357 =$ 5458
b. $6 + 27 + 18 + 5 =$ 56
c. $6.25 + $8.23 + $12.25 =$ $26.73
d. $3 + $2.65 + $0.24 =$ (rewrite $3 as $3.00) $5.89

problem set 1 1. What number is 25 more than 40?   65

2. Johnny had 137 apple seeds in one pocket and 89 in another. He found 9 more seeds in his cuff. How many seeds did he have in all?   235 seeds

3. What is the sum of 876, 950, and 1218?   3044

4. What is 387 plus 96?   483

*5. Monica had $5.22 and earned $1.15 more. How much money did Monica have in all? $6.37

6. The hamburger cost $1.25, the fries cost $0.70, and the drink cost $0.60. What was the total price of the lunch?
$2.55

7.	63	**8.**	632	**9.**	78	**10.**	432
	47		57		9		579
	$+50$		$+198$		$+967$		$+3604$
	160		887		1054		4615

*11. $345 + 67 =$ 412

12. $678 + 4163 =$ 4841

13. $3764 + 96 =$ 3860

14. $875 + 1086 + 980 =$ 2941

15. $10 + 156 + 8 + 27 =$ 201

16.	$3.47	**17.**	$24.15	**18.**	$0.75	**19.**	$0.12
	$+$0.92		$+$ 1.45		$+$0.75		$0.46
	$4.39		$25.60		$1.50		$+$0.50
							$1.08

20. $3.75 + $9.28 =$ $13.03

21. $5 + $1.25 =$ $6.25

22. $3 + $2.57 + $0.68 =$ $6.25

23. $3 + \square = 10$ 7

24. $23 + \square = 100$ 77

25. $56 + \square = 72$ 16

LESSON 2

Subtracting Whole Numbers, Subtracting Decimal Numbers and Money

subtracting   To remove one amount from another amount is to **subtract**. The result of subtraction is called the **difference**.

We can only subtract like things. We can only subtract digits with the same place value. When subtracting we must be sure that digits with the same place value are aligned. We must also be sure to set up the subtraction correctly because order matters in subtraction: $2 - 4$ is not the same as $4 - 2$. When setting up a subtraction problem the first number goes on top.

$$5 - 3 = \begin{array}{r} 5 \\ -3 \\ \hline \end{array}$$

example 2.1 $345 - 67 =$

solution   As with addition we must align place values when we subtract. When we subtract whole numbers, we align the last digits. We subtract the bottom numbers from the top numbers. We "borrow" when it is necessary.

$$\begin{array}{r} {\scriptstyle 2\ \ 13} \\ {\scriptstyle 3\ \ 4^1 5} \\ -\ \ \ 6\ 7 \\ \hline \mathbf{2\ 7\ 8} \end{array}$$

example 2.2   Jim spent $1.25 for a hamburger. If he paid for it with a five-dollar bill, how much change should he get back?

solution   To find out how much of his money is left we must subtract. Order matters when we subtract. The starting amount is put on top. We line up the decimal points to line up place values. Then we subtract.

$$\begin{array}{r} {\scriptstyle 4\ \ \ \ \ 9} \\ \$\ 5\ .\ 0^1 0 \\ -\ \$\ 1\ .\ 2\ 5 \\ \hline \mathbf{\$\ 3\ .\ 7\ 5} \end{array}$$

checking   We may check the answer to a subtraction problem by adding. If we add the answer (difference) to the amount subtracted, the total should equal the starting amount. Using our two examples,

$$\begin{array}{r} 278 \\ +\ 67 \\ \hline 345 \end{array} \text{ check} \qquad \begin{array}{r} \$3.75 \\ +\$1.25 \\ \hline \$5.00 \end{array} \text{ check}$$

We may use subtraction to find missing addends.

$$23 + \boxed{} = 91 \qquad 91 - 23 = \mathbf{68}$$

practice   **a.** $5374 - 168 =$ 5206

            b. $4000 - 2185 =$ 1815

            c. $\$6.00 - \$1.35 =$ \$4.65

            d. $375 + \square = 512$ 137

problem
set 2

1. What number is 27 less than 110?   83

2. What is the difference of 97 and 79?   18

3. How much more than 17 is 50?   33

4. What is 1000 minus 36?   964

5. What is the sum of 386, 98, and 1734?   2218

*6. Jim spent $1.25 for a hamburger. If he paid for it with a five-dollar bill, how much change should he get back?
$3.75

7. Luke wants to buy a $70.00 radio for his car. He has $47.50. How much more does he need? $22.50

8. The bat cost $8.96. The ball cost $3.79. What did the ball and bat cost together? $12.75

9. $\begin{array}{r} 312 \\ -\ 86 \\ \hline 226 \end{array}$	10. $\begin{array}{r} 4106 \\ -1398 \\ \hline 2708 \end{array}$	11. $\begin{array}{r} 4000 \\ -1357 \\ \hline 2643 \end{array}$	12. $\begin{array}{r} 1000 \\ -\ 283 \\ \hline 717 \end{array}$

*13. $345 - 67 =$ 278   14. $351 - 315 =$ 36   15. $3010 - 197 =$
                                                         2813

16. $3,486 + 27,519 =$ 31,005   17. $526 + 87 + 1306 + 919 =$
                                                               2838

18. $\begin{array}{r} \$0.87 \\ -\$0.78 \\ \hline \$0.09 \end{array}$	19. $\begin{array}{r} \$5.00 \\ -\$1.32 \\ \hline \$3.68 \end{array}$	20. $\begin{array}{r} \$0.67 \\ +\$0.76 \\ \hline \$1.43 \end{array}$

21. $\$10 - \$3.75 =$ $6.25   22. $\$3 + \$1.32 + \$0.75 =$ $5.07

23. $27 + \square = 36$ 9   24. $42 + \square = 89$ 47

25. $125 + \square = 500$ 375

LESSON
3

Multiplying Whole Numbers, Multiplying Decimal Numbers and Money

When we add the same number several times, we get the sum. We can get the same answer by multiplying

$$67 + 67 + 67 + 67 + 67 = 335 \qquad \text{and} \qquad 67 \times 5 = 335$$

We see that when a number is added repeatedly, we may multiply and get the same result. Later we will find other uses for multiplying.

Numbers which are multiplied together are called **factors**. The answer is called the **product**.

When multiplying by more than a one-digit number we should remember that we will be multiplying by a number in the tens place or hundreds place or higher place. Multiplying by these places requires us to shift our partial products one, two, or more places to the left as we multiply. We may write zeros in these places. Here the zero shows that 28 has been multiplied by 10, not 1.

$$
\begin{array}{rl}
28 & \text{factor} \\
\times\ 14 & \text{factor} \\
\hline
112 & \\
280 & \\
\hline
392 & \text{product}
\end{array}
$$

When multiplying dollars and cents by a whole number the answer will have cents places, that is, two places after the decimal point.

$$
\begin{array}{r}
\$1.35 \\
\times\qquad 6 \\
\hline
\$8.10
\end{array}
$$

Order does not matter when numbers are multiplied; 4 × 2 = 2 × 4. One way to check multiplication is to reverse the factors and multiply again.

$$
\begin{array}{r}
23 \\
\times\ 14 \\
\hline
92 \\
230 \\
\hline
322
\end{array}
\qquad
\begin{array}{r}
14 \\
\times\ 23 \\
\hline
42 \\
280 \\
\hline
322
\end{array}
\quad
\begin{array}{l}
\text{factors reversed} \\[3.2em]
\text{check}
\end{array}
$$

practice
(See Practice Set A in the Appendix.)

a. 37 × 20 = 740
b. 37 × 23 = 851
c. 46 × 64 = 2944

d. 365 × 56 = 20,440
e. 407 × 37 = 15,059
f. 6 × ☐ = 48   8

problem set 3

1. What is the product of 25 and 12?   300

2. What is the sum of 25 and 12?   37

3. What is the difference of 25 and 12?   13

4. If each of the 31 students brings in 75 aluminum cans, then how many cans will the class collect?   2325 cans

5. What is the total price of one dozen pepperoni pizzas at $7.85 each? $94.20

6. The basketball team scored 63 of its 102 points in the first half of the game. How many points did the team score in the second half?   39 points

7.
$$\begin{array}{r} 368 \\ \times\ 9 \\ \hline 3312 \end{array}$$

8.
$$\begin{array}{r} 407 \\ \times\ 80 \\ \hline 32{,}560 \end{array}$$

9.
$$\begin{array}{r} 28 \\ \times 14 \\ \hline 392 \end{array}$$

10.
$$\begin{array}{r} 370 \\ \times 140 \\ \hline 51{,}800 \end{array}$$

11. 100 × 100 = 10,000

12. 134 × 135 = 18,090

13. 3 × 4 × 5 = 60

14.
$$\begin{array}{r} 3627 \\ 598 \\ +4881 \\ \hline 9106 \end{array}$$

15.
$$\begin{array}{r} 5010 \\ -1376 \\ \hline 3634 \end{array}$$

16.
$$\begin{array}{r} 1000 \\ -\ 26 \\ \hline 974 \end{array}$$

17. $47 + 47 + 47 + 47 + 47 + 47 =$ 282

18. $\$1.36 + \$1.36 + \$1.36 + \$1.36 =$ $5.44

19.	$\$6.35$	**20.**	$\$5.00$	***21.**	$\$1.35$	**22.**	$\$0.27$
	$+\$8.27$		$-\$0.96$		$\times\quad 6$		$\times\quad 10$
	$\$14.62$		$\$4.04$		$\$8.10$		$\$2.70$

23. $37 + \square = 73$ 36

24. $256 + \square = 1000$ 744

25. $\$1.25 + \square = \$8.10$ $6.85

LESSON 4 — Dividing Whole Numbers and Money

dividing When a number is to be separated into a certain number of equal parts, we divide. We can indicate division in several ways. Both of the following notations tell us to divide 24 by 2.

$$24 \div 2 \qquad 2\,\overline{)\,24}$$

example 4.1 Five dozen cookies are to be divided evenly among 15 children. How many cookies should each receive?

solution First we find that 5 dozen cookies is 60 cookies ($5 \times 12 = 60$). Then we divide 60 cookies into 15 parts and find that the number of cookies each child should receive is **4**.

$$15\,\overline{)\,60}\quad \begin{matrix}4\end{matrix}$$

We also divide when we want to separate the members of a group into smaller groups.

example 4.2 How many 14-player soccer teams can be formed if 294 players signed up for soccer?

solution We are to take 294 players and separate them into teams so that there are 14 players on each team. By dividing we find

$$14\,\overline{)\,294}\quad \begin{matrix}21\end{matrix}$$

that the number of teams which can be formed is **21**.

Often the amount which is being divided does not divide evenly. If 30 players are divided into 4 teams there will be 7 players on each team and 2 extra players. For now we will end uneven division by writing a remainder.

$$\begin{array}{r} 7 \\ 4\overline{)30} \\ 28 \\ \hline 2 \end{array} = 7\ r2$$

money   When dividing an amount of dollars and cents, there will be cents in the answer. When money is divided evenly into a whole number of parts, the result is an amount of money which is written with a decimal point directly up from the decimal point in the division box.

$$\begin{array}{r} \$1.60 \\ 3\overline{)\$4.80} \end{array}$$

practice   a. $3\overline{)6042}$ 2014

(See Practice   b. $12\overline{)200}$ 16 r8

Set B in the
Appendix.)

c. $5\overline{)\$8.40}$ $1.68

d. $10\overline{)\$21.30}$ $2.13

problem   *1. Five dozen cookies are to be divided evenly among 15
set 4   children. How many should each receive?   4 cookies

2. If 100 pennies are separated into 4 equal piles, how many pennies will be in each pile?   25 pennies

3. If 100 pennies are put into stacks of 5 pennies each, how many stacks will be formed?   20 stacks

*4. How many 14-player soccer teams can be formed if 294 players signed up for soccer?   21 teams

5. Tom is reading a 280-page book. If he has just finished page 156, how many pages does he still have to read?
124 pages

6. Each month Bill earns $0.75 per customer for delivering newspapers. How much money would he earn in a month in which he had 42 customers? $31.50

7. $6 \overline{)1236}$ **8.** $8 \overline{)5760}$ **9.** $7 \overline{)100}$ **10.** $9 \overline{)2000}$
206   720   14 r2   222 r2

11. $460 \div 20 =$ 23   **12.** $375 \div 25 =$ 15   **13.** $526 \div 18 =$ 29 r4

14.   3629   **15.**   5614   **16.**   2000   **17.**   328
          4716 -4651 -1306 $\times\ 15$
         $+8957$ _____   _____   _____
        _____   963   694   4920
         17,302

18. $563 + 563 + 563 + 563 =$ **19.** $375 \times 16 =$ 6000
2252

20. $480 \times 60 =$ 28,800   **21.** $3 \overline{)\$4.80}$ \$1.60

22. $10 \overline{)\$26.50}$ \$2.65   **23.**   \$3.75   \$33.75
                                        $\times\qquad 9$
                                       _____

24. $\$3 + \$2.86 + \$0.98 =$ **25.** $\$10 - \$6.43 =$ \$3.57
\$6.84

LESSON 5

Paritheses

One use of **parentheses** in mathematics is to show which arithmetic step should be taken first. To find the value of $20 - (10 - 6)$, we first find the value of the amount in parentheses; then we use that answer to do the next step.

$$20 - (10 - 6) \quad \text{problem}$$
$$20 - 4 \quad \text{simplified}$$
$$16 \quad \text{subtracted}$$

We find that when the steps are all addition or all multiplication, parentheses do not affect the value.

$10 + (6 + 3)$ has the same answer as $(10 + 6) + 3$
$3 \times (4 \times 5)$ has the same answer as $(3 \times 4) \times 5$

However, parentheses do affect the value when subtraction and division are involved or when addition and multiplication are mixed.

example 5.1 $5 + (3 \times 4) =$

solution   **We first find the value of the amount in the parentheses; then we use that answer for the next step.**

$$5 + (3 \times 4) \quad \text{problem}$$
$$5 + 12 \quad \text{simplified}$$
$$\mathbf{17} \quad \text{added}$$

example 5.2 $(5 + 3) \times 4 =$

solution   This problem has the same numbers in the same order as the problem above. The $+$ and $\times$ are also in the same order. The answer is different because this time we add first and then multiply.

$$(5 + 3) \times 4 \quad \text{problem}$$
$$8 \times 4 \quad \text{simplified}$$
$$\mathbf{32} \quad \text{multiplied}$$

practice

a. $16 - (3 + 4) = 9$

b. $(16 - 3) + 4 = 17$

c. $24 \div (6 \div 2) = 8$

d. $(24 \div 6) \div 2 = 2$

e. $8 + (6 + 5) = 19$

f. $(8 + 6) + 5 = 19$

g. $9 \times (3 + 4) = 63$

h. $(9 \times 3) + 4 = 31$

problem set 5

1. Jack paid $5 for a hamburger which cost $1.25 and a drink which cost $0.60. How much change should he get back? $3.15

2. In one day the elephant ate 82 pounds of straw, 8 pounds of apples, and 12 pounds of peanuts. How many pounds of food did it eat in all?   102 lb.

3. What number is 25 less than 110?   85

4. The 11 team members were asked to sell raffle tickets at $1 each. If 1 member sold 12 tickets, 6 members sold 10 tickets each, and 4 members sold 8 tickets each, then how much money did the team raise? $104

5. What number must be added to 149 to total 516?   367

6. Judy planned to read a 235-page book in 5 days. How many pages should she read each day? 47 pages

*7. $5 + (3 \times 4) =$ 17

*8. $(5 + 3) \times 4 =$ 32

9. $800 - (450 - 125) =$ 475

10. $600 \div (20 \div 5) =$ 150

11. $(800 - 450) - 125 =$ 225

12. $(600 \div 20) \div 5 =$ 6

13. $144 \div (8 \times 6) =$ 3

14. $(144 \div 8) \times 6 =$ 108

15. $\$5 - (\$1.25 + \$0.60) =$ \$3.15

16. $436 + 436 + 436 + 436 + 436 =$ 2180

17. $\$0.78 + \$0.78 + \$0.78 + \$0.78 =$ \$3.12

18. $24 \overline{)\,1000}$
41 r16

19.
$$\begin{array}{r} 378 \\ \times\ 64 \\ \hline 24{,}192 \end{array}$$

20.
$$\begin{array}{r} 506 \\ \times 370 \\ \hline 187{,}220 \end{array}$$

21.
$$\begin{array}{r} 1010 \\ -\ 989 \\ \hline 21 \end{array}$$

22. $\$42.40 \div 8 =$ \$5.30

23. $\$8 - \$1.75 =$ \$6.25

24. $56 + \square = 432$ 376

25. $8 \times \square = 48$ 6

LESSON
6

Fractional Parts

When we first began to learn about numbers as young children we counted objects. When we count we are using whole numbers. As we grew older we discovered that there are parts of wholes—like parts of a candy bar—which cannot be named with whole numbers. We can name these parts with **fractions.**

We see that this whole "pie" has been cut into **4** parts; **1** part is shaded. The fraction of the pie which is shaded is 1 out of 4 parts. We call this part one-fourth and write it as $\frac{1}{4}$.

example 6.1 What fraction of the "pie" is shaded?

solution The "pie" has been divided into 6 equal parts. We use 6 for the bottom of the fraction. One of the parts is shaded, so we use 1 for the top of the fraction. The fraction of the pie which is shaded is $\frac{1}{6}$.

We can also use fractions to name part of a group. There are 6 members in this group. We can divide this group in half by dividing it into two equal groups with 3 in each half. We write that $\frac{1}{2}$ of 6 is 3.

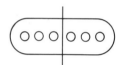

example 6.2 What is $\dfrac{1}{2}$ of 450?

solution To find one-half of 450 we divide 450 into two equal parts and find the amount in one of the parts. We find that $\frac{1}{2}$ of 450 is **225**.

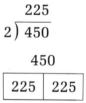

practice What fraction is shaded?

a.  $\dfrac{3}{4}$

b. 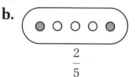 $\dfrac{2}{5}$

c. 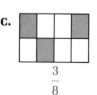 $\dfrac{3}{8}$

d. What is $\dfrac{1}{2}$ of 72?   36

e. What is $\dfrac{1}{2}$ of 1000?   500

f. What is $\dfrac{1}{2}$ of 180?   90

problem set 6

*1. What is $\frac{1}{2}$ of 450?   225

2. What is $\frac{1}{2}$ of 540?   270

3. In four days of sightseeing the Richmonds drove 346 miles, 417 miles, 289 miles, and 360 miles. How many miles did they drive in all?   1412 miles

4. How many hours are there from 6:00 A.M. to 7:00 P.M.?   13 hours

5. How many days are in 52 weeks?   364 days

6. How many $20 bills would it take to make $1000?   50

*7. What fraction of the circle is shaded?

$\frac{1}{6}$

8.
```
  3604
  5186
+ 7145
------
 15,935
```

9.
```
  3001
- 1589
------
  1412
```

10.
```
   376
 ×  87
------
 32,712
```

11.
```
   470
 × 203
------
 95,410
```

12. $20 − $11.98 = $8.02

13. 596 − (400 − 129) = 325

14. 357 ÷ (84 ÷ 4) = 17

15. 8) 4016   502

16. 15) 6009   400 r9

17. 36) 9000   250

18. 76 + 76 + 76 + 76 + 76 + 76 + 76 = 532

19. 532 ÷ 7 = 76

20. $6.35 × 12 = $76.20

21. $6.35 ÷ 5 = $1.27

22. $3.46 + $12 = $15.46

23. 365 + ☐ = 653   288

24. 9 × ☐ = 72   8

25. What fraction of the square is shaded?

$\frac{1}{4}$

LESSON
7

Linear Measure

As civilized people we have agreed upon certain units of measure. In the United States we have two systems of units that we use to measure length. One system is the **U.S. Customary system**. Some of the units in this system are inches, feet, yards, and miles. The other system is the **metric system**, or SI. Some of the units in the metric system are millimeters, centimeters, meters, and kilometers. In this book we will concentrate on lengths measured in inches, millimeters, and centimeters.

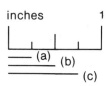

When we measure using inch units we often need to use fractions to name the distances we measure. Inches are commonly divided into halves and quarters. This inch is divided into 4 equal parts. Line (a) is 1 part long so line (a) is $\frac{1}{4}$ inch long. Line (b) is 2 parts long and is $\frac{2}{4}$ inch. This is the same as $\frac{1}{2}$ inch. Line (c) is 3 parts long so line (c) is $\frac{3}{4}$ inch long.

example 7.1   How long is the line segment?

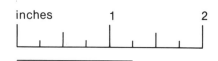

solution   The line is one whole inch plus a fraction. The fraction is one of four parts: $\frac{1}{4}$. The length of the line is **$1\frac{1}{4}$ in**.

example 7.2   How long is the line segment?

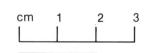

solution   We simply read the scale to see that the line is **2 cm** long.

Note: The abbreviation for inches is ended with a period. The abbreviations for other units are not ended with a period.

practice   How long is each line segment?

a.

$1\frac{3}{4}$ in.

b.

25 mm

problem set 7

1. To earn money for gifts Debbie sold decorated pine cones. If she sold 100 cones at $0.25 each, how much money did she earn? $25.00

2. There are 365 days in a normal year. April 1 is the 91st day. How many days are left in a year after April 1?   274

3. The Smiths are planning to complete a 1890 mile trip in 3 days. If they drive 596 miles the first day and 612 miles the second day, how far must they travel the third day?
682 miles

4. On July 30 the sun rose at 5:57 A.M. and set at 7:57 P.M. There were how many hours between sunrise and sunset?
14 hours

5. What is $\frac{1}{2}$ of 234?   117

6. What fraction is shaded? $\frac{3}{8}$

7.
```
  3654
  2893
+ 5614
------
 12,161
```

8.
```
  4101
− 1376
------
  2725
```

9.
```
   28
 × 74
-----
 2072
```

10.
```
   906
 ×  47
------
 42,582
```

11. 6) 5000   833 r2

12. 23) 800   34 r18

13. 60) 3174   52 r54

14. 3675 + 287 + 37,248 =
41,210

15. $10 + $8.75 = $18.75

16. $10 − $8.75 = $1.25

17.
```
  $4.32
 ×   20
------
 $86.40
```

18.
```
  $0.48
 ×   24
------
 $11.52
```

19. $8.75 ÷ 25 =
$0.35

20. 6) ⬚   42
(7)

21. 58 + ⬚ = 213
155

22. 6 × ⬚ = 96
16

23. Which of these numbers is greatest?   312   321   123
321

*24. How long is the line segment?

$1\frac{1}{4}$ in.

*25. How long is the line segment?

2 cm

LESSON 8

Perimeter

The distance around a shape is called its **perimeter**. The perimeter of this square is the distance around it. If we were to lay a string on the sides of the square so that the string exactly reached around the square, then the length of the string would equal the perimeter of the square.

We use units like inches or centimeters to measure distances. Here we have a rectangle which is 3 cm long and 2 cm wide. If we were to trace the perimeter of the rectangle, our pencil would travel 3 cm then 2 cm then 3 cm then 2 cm to get all the way around the rectangle. Thus,

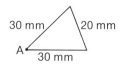

Perimeter = 3 cm + 2 cm + 3 cm + 2 cm = **10 cm**.

example 8.1   What is the perimeter of the triangle?

30 mm   20 mm

A   30 mm

solution   The perimeter of a shape is the distance around it. If we use a pencil to trace this triangle from point A, the point of the pencil would travel 30 mm then 20 mm then 30 mm. Adding these distances, we find that the perimeter is **80 mm**.

practice   What is the perimeter of each shape?

a. Square 48 mm   c. Rectangle 70 mm   e. Pentagon 5 cm

12 mm

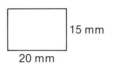
15 mm
20 mm

1 cm 1 cm
1 cm   1 cm
1 cm

b. Equilateral   d. Trapezoid   55 mm
triangle 6 cm

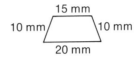
15 mm
10 mm   10 mm
20 mm

2 cm

problem set 8

1. In an auditorium there are 25 rows of chairs with 18 chairs in each row. How many chairs are in the auditorium? 450 chairs

2. How many years was it from 1903 to 1969? 66 years

3. What time is 6 hours after 9:00 A.M.? 3 P.M.

4. Robin Hood divided 140 of his merry men into 5 equal groups. How many were in each group? 28 men

*5. What is the perimeter of the triangle? 80 mm

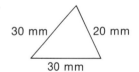
30 mm   20 mm
30 mm

6. What is $\frac{1}{2}$ of 654? 327

7. What fraction of the rectangle is shaded?

$\frac{3}{10}$

8. $4 \overline{)900}$ 225

9. $10 \overline{)373}$ 37 r3

10. $12 \overline{)1500}$ 125

11. $39 \overline{)800}$ 20 r20

12. $300 \times 200 =$ 60,000   **13.** $625 \times 24 =$ 15,000

14. $4629 + 96 + 387 =$ 5112   **15.** $\$3.84 + \$0.83 + \$12 =$
                                              $16.67

16. $3106 - 248 =$ **17.** $\$1 - \$0.72 =$ **18.** $\$5 - \$1.48 =$
  2858 $0.28 $3.52

19. $825 \div 8 =$ **20.** $2410 \div 12 =$ **21.** $1234 \div 15 =$
  103 r1   200 r10   82 r4

22. $315 + \square = 397$ 82   **23.** $5 \times \square = 85$ 17

24. How long is the line segment?

inches   1   2 $1\frac{1}{2}$ in.

25. How long is the line segment?

mm   10   20   30   40   27 mm

LESSON
9

Comparing, Sequences, Odd-Even

comparison symbols
When we are asked to **compare** two numbers we must state that the numbers are equal or that one is greater and the other is less. We may show these relationships with comparison symbols. If the numbers are equal, the comparison sign we use is the **equals** sign (=).

$$1 + 1 = 2$$

If the numbers are not equal we use the **greater than/less than** sign (>). The greater than/less than sign may point to the right or to the left (> or <). The symbol < has a small end that is read as "is less than" and a large end that is read as "is greater than." We read the end we come to first and do not read the other end. We may read a comparison in either direction.

$\longrightarrow$ Reading from left to right $\longleftarrow$ Reading from right to left
$2 < 4$ "2 is less than 4" $2 < 4$ "4 is greater than 2"

When we write a greater than/less than symbol between two numbers we write it so that the small end points to the smaller number and the large end points to the larger number.

example 9.1   Compare: 5012 $\bigcirc$ 5102

solution   In place of the circle we should write $=$ or $>$ or $<$ to make the statement true. Since 5012 is less than 5102, we let the small end point to the 5012.

$$5012 < 5102$$

sequences   A **sequence** is an ordered list of numbers that follows a certain rule.

(a) 5, 10, 15, 20, 25, . . .
(b) 5, 10, 20, 40, 80, . . .

Sequence (a) is an **addition sequence** because the same number is added to each term of the sequence to get the next term. In this case 5 is added to each term. Sequence (b) is a **multiplication sequence** because each term of the sequence is multiplied by the same number to get the next term. In (b) each term is multiplied by 2. When we are asked to find a missing number in a sequence, we must inspect the numbers to discover the rule for the sequence. We will then be able to use the rule to find other numbers in the sequence.

example 9.2   What is the next number in this sequence? 1, 3, 9, 27, ___

solution   Inspecting the numbers, we find that each term in the sequence can be found by multiplying the term before it by 3. Multiplying 27 by 3 we find that the next term in the sequence is **81**.

odd-even   The numbers . . . 0, 2, 4, 6, 8, . . . form a special sequence called **even numbers**. We say the even numbers when we "count by twos." The whole numbers which are not even numbers are **odd numbers**. The odd numbers are . . . 1, 3, 5, 7, 9, ...

practice   Compare:

a. $2 + 2 + 2 + 2 \bigcirc 4 \times 2$ =   **d.** $2 + 2 + 3 \bigcirc 3 + 2 + 2$ =

b. $3 \times 2 \bigcirc 2 + 2$ >   **e.** $4 \times 3 \bigcirc 24 \div 2$ =

c. $2 \times 3 \bigcirc 2 + 3$ >   **f.** $1 \times 1 \times 1 \bigcirc 1 + 1 + 1$ <

problem set 9

*1. What is the next number in this sequence? 1, 3, 9, 27, _81_

2. How many years was it from 1492 to 1603?   111 years

3. What time is 5 hours before 1:00 P.M.?   8 A.M.

4. Martin is carrying groceries in from the car. If he can carry 2 bags at a time, how many trips will it take him to carry in 9 bags?   5 trips

5. What is the perimeter of the rectangle?   60 mm

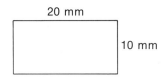

20 mm

10 mm

6. What is $\frac{1}{2}$ of \$5.80?   \$2.90

7. What fraction of the triangle is shaded? $\frac{1}{4}$

*8. Compare: $5012 \bigcirc 5102$.   <

9. Compare: $1 \bigcirc \frac{1}{2}$.   >

10. What number is next in this sequence? 16, 18, 20, 22, _24_
24

11.   478   **12.**   5000   **13.**   420   **14.**   78
      3692 -3176 $\times\ 60$ $\times 36$
   $+\ \ \ 45$ ————   ————   ————
   ————   1824   25,200   2808
      4215

15. $9\overline{)7227}$ 803   **16.** $25\overline{)7600}$ 304   **17.** $20\overline{)8014}$ 400 r14

18. $7136 \div 100$ 71 r36   **19.** $736 + 2875 + 93 = 3704$

20. $6.35 + $0.90 + $8 =
$15.25

21. $20 − $8.42 = $11.58

22. $3.14 × 12 = $37.68

23. $25.20 ÷ 6 = 4.2

24. How long is the line segment?

inches   1   2

$\frac{3}{4}$ in.

25. How long is the line segment?

cm   1   2   3   4   5

4 cm

LESSON 10

Reading Scales and Graphs

scales   Numerical information is often presented to us in the form of a **scale** or **graph**. A scale is a display of numbers with an indicator to show where a certain measure falls on the scale. **The trick to reading a scale is to discover the value of the marks on the scale.** Marks on a scale may show every unit or only every two, five, ten, or another number of units. We must study the scale to find the value of the units before we try to read the indicated number.

example 10.1   What temperature is shown on the thermometer?

solution   As we look at the temperature scale on the thermometer we see that there are 5 marks from 0° to 10°. Each mark then must equal 2°. The indicator falls two marks above zero. With each mark equal to 2° the temperature shown is **4°**.

graphs   A graph is a picture which helps us visualize comparisons or changes. Graphs use lines, bars, pictures, or parts of circles to illustrate amounts. As with scales it is necessary to discover the value of the marks on the graph before we attempt to read a value from a graph.

Price of bananas

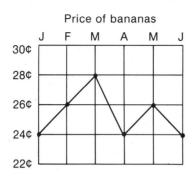
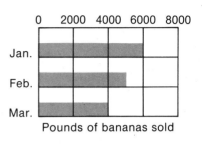

Pounds of bananas sold

The first graph is a line graph which shows the price of one pound of bananas during the first six months of the year. The second graph is a bar graph which shows the number of pounds of bananas sold during the first three months of the year.

practice   Use the graphs to answer these questions.

a. What was the price of a pound of bananas in May?   26¢

b. The price of bananas was highest during what month?
March

c. During which month were only 4000 pounds of bananas sold?   March

d. How many pounds of bananas were sold in February (approximately)?   About 5000 pounds

**problem
set 10**

1. Tickets for the movie cost $4.75 for adults and $2.50 for children. What is the total cost of the tickets for two adults and three children? $17.00

2. How many years was it from 1620 to 1776?   156 years

3. What time is 13 hours after 3:00 P.M?   4 A.M.

*4. What temperature is shown on the thermometer?   4°

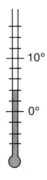

5. What is the perimeter of the square?
    40 mm

6. What is $\frac{1}{2}$ of $6.50? $3.25

7. Compare: $3 \times 2 \bigcirc 3 + 2$ >

8. What fraction of the circle is **not** shaded?
    $\frac{3}{4}$

9. What number is 562 less than 10,000?   9438

10. 365	**11.** 146	**12.** 78	**13.** 907
$\times 100$	$\times 240$	$\times 48$	$\times 36$
36,500	35,040	3744	32,652

14. $10\overline{)4260}$ 426   **15.** $20\overline{)4260}$ 213   **16.** $15\overline{)4260}$ 284

17. $5637 + 28{,}415 + 3716 + 975 =$ 38,743

18. $28{,}347 - 9{,}637 =$ 18,710   **19.** $8 + $3.49 =$ $11.49

20. $10 - $0.75 =$ **21.** $0.56 \times 60 =$ **22.** $6.20 \div 4 =$
    $9.25 $33.60 $1.55

Use the graph to answer questions 23, 24, and 25.

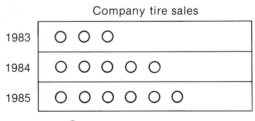

Company tire sales

○ Represents 1000 tires

23. Each ○ means how many tires?   1000 tires

24. How many tires were sold in 1984?   5000 tires

25. How many **more** tires were sold in 1985 than in 1983?
    3000 tires

LESSON
11

Operations of Arithmetic

The **operations** of arithmetic are addition, subtraction, multiplication, and division. There are mathematical terms for the answers we get when we perform these operations.

Sum	The answer when you add
Difference	The answer when you subtract
Product	The answer when you multiply
Quotient	The answer when you divide

Throughout these lessons you will be asked to find the **sum**, **difference**, **product**, and **quotient** of various kinds of numbers.

example 11.1   What is the sum of 8 and 4?

solution   The sum is the answer when you add. When you add 8 and 4, the sum is **12**.

example 11.2   What is the difference between the product of 8 and 4 and the sum of 8 and 4?

solution   We are asked to find the difference, so we must subtract. The product of 8 and 4 is 32. The sum of 8 and 4 is 12. If we subtract 12 from 32, we get a difference of **20**.

$$\begin{array}{r} 32 \\ -\ 12 \\ \hline 20 \end{array}$$ product of 8 and 4
sum of 8 and 4
difference

practice   a. What is the sum of 9 and 3?   12
b. What is the difference of 9 and 3?   6
c. What is the product of 9 and 3?   27
d. When the sum of 9 and 3 is divided by the difference of 9 and 3, what is the quotient?   2

problem set 11

*1. What is the sum of 8 and 4?   12

*2. What is the difference between the product of 8 and 4 and the sum of 8 and 4?   20

3. If the product of 6 and 4 is divided by the difference of 8 and 5, what will be the quotient?   8

4. Your heart beats about 72 times per minute. At that rate, how many times will it beat in one hour?   4320

5. When Jack went to bed at night, the beanstalk was one meter tall. When he woke up in the morning, the beanstalk was one thousand meters tall. How many meters had the beanstalk grown during the night?   999 m

6. $0.65 + $0.40 =
   $1.05
   7. $7 + \square = 15$ 8
   8. $1000 - 386 =$
      614

9. $1000 - (100 - 10) =$ 910
   10. $42 + 596 + 3488 =$ 4126

11. Compare: $15{,}000 \bigcirc 10{,}500$ >

12. $8\overline{)1000}$ 125
    13. $10\overline{)987}$ 98 r7
    14. $12\overline{)420}$ 35

15. $600 \times 300 =$ 180,000
    16. $365 \times \square = 365$ 1

17. What comes next in the sequence? 2, 6, 10, _14_

18. $2 \times 3 \times 4 \times 5 =$ 120
    19. What is $\frac{1}{2}$ of 360?   180

20. Which digit is in the hundreds' place in 123,456?   4

21. What is the product of eight and one hundred twenty-five?
    1000

22. How long is the line segment?
    $2\frac{1}{4}$ in.

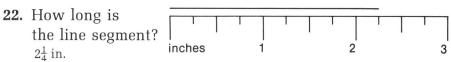
inches   1   2   3

23. What fraction of the circle is shaded?
    $\frac{3}{8}$

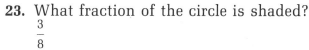

24. What is the perimeter of the square?
36 mm

□ 9 mm

25. What is the sum of the first five odd numbers greater than zero?   25

LESSON 12

Place Value Through Billions

In our number system the value of a digit depends upon its position. The value of each position is called its **place value**.

| Hundred billions | Ten billions | Billions | Hundred millions | Ten millions | Millions | Hundred thousands | Ten thousands | Thousands | Hundreds | Tens | Ones |

example 12.1   In the number 123,456,789,000 which digit is in the ten-millions' place?

solution   Either by counting places or looking at the chart we find that the digit in the ten-millions' place is **5**.

example 12.2   In the number 5,764,283 what is the place value of the digit 4?

solution   By counting places or looking at the chart we can see that the place value of 4 is **thousands**.

practice   **a.** Which digit is in the millions' place in 123,456,789?   3
b. What is the place value of the 3 in 9,876,543?   ones
c. Which digit is in the hundreds' place in 987,654?   6
d. The zero holds what place value in 908,321?   ten thousands
e. In 123,456,789,000 what is the digit in the ten-billions' place?   2
f. What is the place value of the 1 in 1,234,567,890?   billion

problem
set 12

1. What is the difference between the product of 1, 2, and 3 and the sum of 1, 2, and 3?   0

2. The earth is about ninety-three million miles from the sun. Write that number.   93,000,000

3. Jack found two dozen golden eggs. If the value of each egg was $1000, what was the value of all the eggs Jack found? $24,000

4. Robin bought two arrows for $1.75 each. If he paid the good merchant with a five-dollar bill, how much did he receive in change? $1.50

5. What is the perimeter of the rectangle?   56 mm

10 mm

18 mm

6. $6 \times \square = 60$ 10   7. What is $\frac{1}{2}$ of 100?   50

8. Compare: $300 \times 1 \bigcirc 300 \div 1$ =

9. $(3 \times 3) - (3 + 3) =$ 3

10. What is the next number in this sequence? 1, 2, 4, 8, __16__

11. $1 + 23 + 456 =$ 480   12. $1010 - 909 =$ 101

13. $1234 \div 10 =$ 123 r4   14. $1234 \div 12 =$ 102 r10

15. What is the sum of the first five even numbers greater than zero?   30

16. How many millimeters long is the line segment?   32 mm

mm   10   20   30   40

*17. In the number 123,456,789,000 which digit is in the ten-millions' place?   5

*18. In the number 5,764,283 what is the place value of the digit 4?   thousands

19. Which digit is in the hundred-thousands' place in 987,654,321?   6

20. $1 \times 10 \times 100 \times 1000 =$
1,000,000

21. $\$3.75 \times 3 = \$11.25$

22. $22 \times \boxed{} = 0$ 0

23. $100 + 200 + 300 + 400 =$
1000

24. $24 \times 26 = 624$

25. $\boxed{} \overline{)\,568}$ $\overset{71}{}$ 8

LESSON 13

Reading and Writing Whole Numbers

Large numbers are easy to read and write if we use commas to group the digits. To place commas we begin at the right and move to the left, writing a comma after each three digits.

Putting commas in 1234567890 we get 1,234,567,890.

Commas help us read large numbers by marking the end of the billions, millions, and thousands. We need only to read the three-digit number in front of each comma, then say "billion" or "million" or "thousand" when we reach the comma.

billion   million   thousand   no comma

example 13.1   Read 1,000,500,010.

solution   "One **billion**, five hundred **thousand**, ten"
Note: Since there are no millions we don't read the millions' comma.

example 13.2 Read 4005010.

solution First we insert the commas and then read, "Four **million**, five **thousand**, ten."
Note: When writing out numbers, put commas after the words billion, million, thousand.

example 13.3 Use words to write 521.

solution **Five hundred twenty-one**.
Note 1: Do not say or write **and** with whole numbers.
Note 2: Hyphenate all compound numbers from 21 through 99.

example 13.4 Write the numeral for five million, two hundred thousand.

solution We must have a digit in each place after the first digit. We use zeros as place holders in the last three places and write **5,200,000**.

practice
(See Practice
Set C in the
Appendix.)
Use words to write:
a. 10,000,000   ten million
b. 5,010,200,000   five billion, ten million, two hundred thousand
c. 1020300   one million, twenty thousand, three hundred

Use digits to write:
d. Ten million, five hundred thousand   10,500,000
e. One billion, twenty million   1,020,000,000
f. One hundred two thousand, three hundred   102,300

**problem
set 13**

1. When the sum of 8 and 5 is subtracted from the product of 8 and 5, what is the difference?   27

2. The moon is about two hundred fifty thousand miles from the earth. Write that number.   250,000

*3. Use words to write 521.   five hundred twenty-one

*4. Use digits to write five million, two hundred thousand.
5,200,000

5. Robin Hood roamed Sherwood Forest with sevenscore merry men. A score is twenty. How many merry men roamed with Robin?   140

6. The beanstalk was 1000 meters tall. The giant had climbed down 487 meters before Jack could chop down the beanstalk. How far did the giant fall?   513 m

7. $4.56 ÷ 3 =$ $1.52

8. 99 + 100 + 101 =   300

9. 9 × 10 × 11 =   990

10. Which digit is in the thousands' place in 54,321?   4

11. What is the place value of the 1 in 1,234,567,890?   billions

12. What is the perimeter of the equilateral triangle?   54 mm

18 mm

13. 5432 ÷ 100 =
54 r32

14. 60,600 ÷ 30 =
2020

15. 1000 ÷ 7 =
142 r6

16. 500 − ☐ = 460   40

17. Compare. 63 + 36 ◯ 36 + 63   =

18. The sequence in this problem has a rule that is different from the rules for an addition sequence or a multiplication sequence. What is the missing number in the sequence?

1, 4, 3, 6, 5, 8, _7_

19. What is $\frac{1}{2}$ of 5280?
2640

20. 365 ÷ ☐ = 365   1

21. (5 + 6 + 7) ÷ 3 =   6

22. What is the length of the arrow? $1\frac{3}{4}$ in.

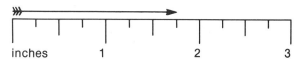

23. $1 + 2 + 3 + 4 + 5 + 6 + 7 + 8 + 9 + 10 =$ 55

24. $25 + 25 + 25 + 25 + 25 + 25 + 25 + 25 + 25 + 25 =$ 250

25. If $5 * 2$ equals 3, and $6 * 2$ equals 4, then $56 * 22$ equals what?   34

LESSON
14

The Number Line

A **number line** can be used to arrange numbers in order.

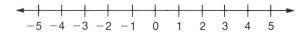

On the number line above, the points to the right of zero represent **positive** numbers. The points to the left of zero represent **negative** numbers. Zero is neither positive nor negative. The vertical marks represent whole amounts (integers). Between the marks are points which represent fractional numbers. We will be learning much more about negative numbers and fractional numbers in later lessons.

example 14.1   What number is three spaces to the left of 2 on the number line? (A space is the distance between two vertical marks.)

solution   Start at 2 and count left three spaces ... 1, 0, **−1**.

example 14.2   What number is halfway between 1 and 5 on the number line?

solution   We see three whole numbers between 1 and 5. They are 2, 3, and 4. The one exactly in the middle is **3**.

example 14.3   What number is halfway between 0 and 1?

solution   There is no whole number between zero and one, but we can see that there are fractions between 0 and 1. The fraction halfway between is $\frac{1}{2}$.

practice   **a.** What number is three spaces to the left of 0? -3
b. What number is four spaces to the right of -2?   2
c. What number is halfway between 0 and 4?   2
d. What number is halfway between -1 and 2? $\frac{1}{2}$

problem set 14

1. What is the quotient when the sum of 15 and 12 is divided by the difference of 15 and 12?   9

2. What is the place value of the 7 in 987,654,321,000?   billions

3. Light travels at a speed of about one hundred eighty-six thousand miles per second. Write that number.   186,000

*4. What number is three spaces to the left of 2 on the number line? -1

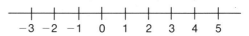

*5. What number is halfway between 1 and 5 on the number line?   3

*6. What number is halfway between 0 and 1 on the number line? $\frac{1}{2}$

7. Little John stood 7 feet tall. Robin Hood was only 5 feet 11 inches tall. How many **inches** taller than Robin was Little John?   13 in.

8. Compare: $1 + 2 + 3 + 4 \bigcirc 1 \times 2 \times 3 \times 4$ $<$

9. What is the perimeter of the right triangle?   60 mm

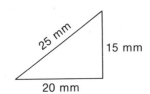

10. What is the next number in the sequence? 16, 8, 4, <u> 2 </u>

11. $365 + \square = 500$ 135   **12.** What is $\frac{1}{2}$ of $(6 + 8)$?   7

13. $1020 \div 100 =$ **14.** $36,180 \div 12 =$ **15.** $564 \div 18 =$
    10 r20   3015   31 r6

16. $1234 + 567 + 89 =$ 1890   **17.** $51,023 - 6,375 =$ 44,648

18. $10 \times 11 \times 12 =$ 1320   **19.** $\$3.05 - \$2.98 =$ \$0.07

20. How many centimeters long is the nail?   4 cm

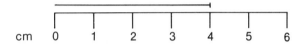

cm   0   1   2   3   4   5   6

21. $100 \times 100 \times 100 =$ 1,000,000

22. What digit is in the ten-thousands' place in 123,456,789?
    5

23. $23 \times 24 = 24 \times \square$ 23

24. $(3 \times 18) - (18 + 18 + 18) =$ 0

25. If the sum of the first five positive odd numbers is subtracted from the sum of the first five positive even numbers, what is the difference?   5

LESSON
15

Negative Numbers

People who live in cold climates become familiar with numbers less than zero at an early age. Imagine that the temperature outside is 2° but is getting colder. The temperature drops one degree, then one more degree, then one more degree. The temperature is now one degree below zero. One degree below zero can be written $-1°$. Read -1 as "negative one." "Negative one" means one less than zero. If you subtract a larger number from a smaller number (like 2 − 3), the answer will be a negative number. One way to find the answer to such questions is to use the number line. Maybe you can figure out a faster way to find the answer.

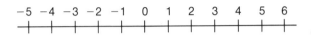

example 15.1   Subtract: 1 − 3 =

solution   Start at 1 and count "back" 3 spaces. You should end up at **−2**.

example 15.2   Subtract 5 from 2.

solution   **Order matters in subtraction**. Start at 2 and count to the left 5 spaces. You should end up at **−3**.

practice   a. $3 - 4 = -1$
    b. $0 - 2 = -2$
    c. $5 - 8 = -3$
    d. $3 - 6 = -3$
    e. $1 - 5 = -4$

    f. Subtract 5 from 4 -1
    g. Take 6 from 2 -4
    h. Solve 2 minus 5 -3
    i. What number is 10 less than 5? -5
    j. Read this number: -14
       "negative 14"

problem set 15

1. What is the product of one hundred nine and eighty-seven?   9483

2. On the Fahrenheit scale of temperature, water freezes at 32° and boils at 212°. How many degrees difference is there between the freezing and boiling points?   180°

3. There are about three hundred twenty little O's of cereal in an ounce. About how many O's are there in a one-pound box? (1 pound = 16 ounces.)   5120

4. There are 31 days in August. How many days are left in August after August 3?   28

*5. Subtract: $1 - 3 = -2$

*6. Subtract 5 from 2. Answer? -3

7. What number is 8 less than 5? -3

8. What number comes next in this sequence? 6, 4, 2, 0, −2, ____
−4

9. What is the temperature reading on this thermometer? −6

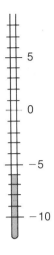

10. $10 − 10¢ = $9.90

11. What is $\frac{1}{2}$ of $3.50? $1.75

12. Which hundred is 587 closest to? 600

13. 9 + 87 + 654 + 3210 =
3960

14. 1000 − 65 = 935

15. 4320 ÷ 9 = 480

16. 36 ⟌ 493   13 r25

17. (8 + 9 + 16) ÷ 3 = 11

18. ☐ × 63 = 63   1

19. 76 ÷ ☐ = 1   76

20. 574 × 76 = 43,624

21. Compare: 100 − 20 ◯ 100 − 19   <

22. There are 10 millimeters in 1 centimeter. How many millimeters long is the paper clip? 30 mm

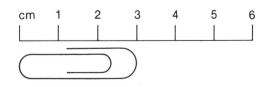

23. 1200 ÷ 300 = 4

24. What is the place value of the 5 in 12,345,678?   thousands

25. Which digit is in the ten-billions' place in 123,456,789,000?
2

LESSON 16

Rounding Whole Numbers and Estimating

rounding   When we **round** a number, we are finding another number it is close to. The number line can help us visualize rounding:

If we are to round 667 to the nearest ten, we can see that 667 is closer to 670 than it is to 660. If we are to round 667 to the nearest hundred, we can see that 667 is closer to 700 than 600.

example 16.1   Round 56,789 to the nearest thousand.

solution   The number we are rounding is between 56,000 and 57,000. It is closer to **57,000**.

example 16.2   Round 550 to the nearest hundred.

solution   The number we are to round is halfway between 500 and 600. When this happens, we round **up** so we get **600**.

estimating   Rounding can help us estimate the answer to a problem. **Estimating** is a quick, easy way to get "close" to the answer. It can help us decide if our exact answer is reasonable, or if we have made a mistake. To estimate we round the numbers **before** we add, subtract, multiply, or divide.

practice
(See Practice
Set D in the
Appendix.)   Use rounded numbers to estimate the answers.

a. $397 + 206 = $ 600

b. $703 - 598 = $ 100

c. $29 \times 31 = $ 900

d. $29 \overline{)591}$ 20

e. $19 \times (41 + 58) = $ 2000

f. $3986 \div 41 = $ 100

problem
set 16

1. What is the difference between the product of 20 and 5 and the sum of 20 and 5?   75

2. Columbus landed in the Americas in 1492. The Pilgrims landed in 1620. How many years after Columbus did the Pilgrims land in America?   128 years

3. Robin Hood separated his 140 merry men into 5 equal groups. One group he sent north, one south, one east, one west. The remaining group stayed in camp. How many merry men stayed in camp?   28

4. Which digit is in the hundred-thousands' place in 159,342,876?   3

5. Write the name of this number: 5,010.   five thousand, ten

6. What number is halfway between 5 and 11 on the number line?   8

*7. Round 56,789 to the nearest thousand.   57,000

*8. Round 550 to the nearest hundred.   600

9. Estimate the product of 295 and 406 by rounding to the nearest 100 before multiplying.   120,000

10. $45 + 5643 + 287 =$ 5975   11. $40,312 - 14,908 =$ 25,404

12. $7308 \div 12 =$ 13. $100 \overline{)5367}$ 14. $(5 + 11) \div 2 = 8$
    609   53 r67

15. What is $\frac{1}{2}$ of $5? $2.50   16. $\square - 30 = 20$ 50

17. $0.25 \times 10 = 2.50 18. $325 \times (324 - 323) =$ 325

19. Compare: $1 + (2 + 3) \bigcirc (1 + 2) + 3$ $=$

20. $1 - 5 = -4$

21. What number is 10 less than 5? -5

22. What comes next in the sequence? 100, 80, 60, 40, <u>20</u>

Use the graph to answer questions 23, 24, and 25.

	Pounds of peanuts eaten daily								
	10	20	30	40	50	60	70	80	90
Dumbo	▓	▓	▓						
Mumbo	▓	▓	▓	▓	▓	▓			
Jumbo	▓	▓	▓	▓	▓	▓	▓	▓	

23. By the graph, how many more pounds of peanuts does Jumbo eat each day than Dumbo?   50 lb.

24. All together, how many pounds do the three elephants eat each day?   170 lb.

25. How many pounds would Mumbo eat in one week? 420 lb.

LESSON 17

The Bar as a Division Symbol

division bar You are familiar with two ways of showing division. Sixty divided by three can be written $60 \div 3$ or $3\overline{)60}$. There is a third way of showing division which you will see more often as you continue to learn math. The third way is with a **division bar**. For example, you may see $\frac{60}{3}$. We read the division problem from top to bottom as "Sixty **divided by** three." You may change the form of the problem to $3\overline{)60}$ to do the division.

division terms Division involves three numbers. The number being divided is called the **dividend**. The number dividing into that number is called the **divisor**, and the answer is called the **quotient**.

$$\text{Divisor}\overline{)\text{dividend}}^{\text{quotient}} \qquad \frac{\text{Dividend}}{\text{Divisor}} = \text{quotient}$$

example 17.1 Divide: $\dfrac{48}{3} =$

solution Read "48 divided by 3." Change form if necessary.

$$3\overline{)48}^{\,16}$$

example 17.2 Divide: $\dfrac{2+4+6}{3}$

solution First find the sum of $2+4+6=12$. Then divide by 3. $3\overline{)12}=4$

example 17.3 Rewrite $2 \div 3$ with a division bar but do not divide.

solution "Two divided by three" is written $\dfrac{2}{3}$.

practice Divide:

a. $\dfrac{16}{4}$ 4

b. $\dfrac{5}{5}$ 1

c. $\dfrac{4+5+6}{3}$ 5

d. $\dfrac{123}{3}$ 41

Write with division bar:

e. $3 \div 4$ $\dfrac{3}{4}$

f. $5\overline{)6}$ $\dfrac{6}{5}$

g. The dividend is 7; the divisor is 8. $\frac{7}{8}$

h. Four divided by five $\dfrac{4}{5}$

problem set 17

1. What is the sum of twelve thousand, five hundred and ten thousand, six hundred ten? 23,110

2. In 1903 the Wright brothers made the first powered airplane flight. In 1969 men first landed on the moon. How many years was it from the first powered airplane flight to the first manned moon landing? 66 years

3. Captain Hook often ran from the sound of ticking clocks. If he could run 72 yards in 12 seconds, how far could he run in 1 second? 6 yd.

4. How many spaces are between −1 and 2 on the number line? ₃

5. Estimate the sum of 5280 and 1760 by rounding to the nearest thousand before adding. ₇₀₀₀

*6. Divide: $\dfrac{48}{3}$ = ₁₆

*7. Divide: $\dfrac{2 + 4 + 6}{3}$ = ₄

8. $4 - 6$ = ₋₂

*9. Rewrite $2 \div 3$ with a division bar but do not divide. $\dfrac{2}{3}$

10. A square has sides 10 cm long. What is its perimeter? 40 cm

11. How long is the line segment? $2\frac{1}{2}$ in.

12. $3 − $1.25 = $₁.₇₅

13. $365 + 4576 + 50{,}287$ = 55,228

14. $19 \times 20 \times 21$ = ₇₉₈₀

15. Compare: $19 \times 21 \bigcirc 20 \times 20$ <

16. $5280 \div 44$ = ₁₂₀

17. What number is missing in this sequence? 5, 10, _15_, 20, 25

18. Which digit is in the hundred-millions' place in 987,654,321? ₉

19. $250{,}000 \div 100$ = ₂₅₀₀

20. 3.75×10 = $₃₇.₅₀

21. $16 + 14 = 14 + \square$ ₁₆

22. The magician pulled 38 rabbits out of his hat. One-half of the rabbits were white. How many were not white? ₁₉

23. $100 - (50 - 25) =$ ₇₅   **24.** $\square - 20 = 30$ ₅₀

25. What is the sum of the first six positive odd numbers?
₃₆

LESSON
18

Average

The **average** of two numbers is the number halfway between them. On a number line the average is halfway between the two numbers. The average of 6 and 8 is 7. What is the average of 6 and 10?

The average of three or more numbers is still a central number. The average will be more than the smallest number and less than the largest, but the average won't be the middle number unless the numbers are equally spaced. To find the average we can **add** the numbers then **divide** the sum by the number of numbers we are averaging. Step 1: Add the numbers. Step 2: Divide by the number of numbers. This two-step method of **adding** and **dividing** will always give us the average.

example 18.1   Find the average of 3, 7, and 8.

solution   There are three numbers. Add the numbers and divide by 3. With a division bar it looks like this: $\dfrac{3 + 7 + 8}{3} = \dfrac{18}{3} = \mathbf{6}$. The average of 3, 7, and 8 is **6**.

example 18.2   What number is halfway between 27 and 81?

solution   We could use the number line or find the average. With a division bar it looks like this: $\dfrac{27 + 81}{2} = \dfrac{108}{2} = 54$. The average of 27 and 81 is **54**.

practice
(See Practice
Set E in the
Appendix.)

Add and divide to find the average.
a. 3, 8, 10   7
b. 12, 14, 22   16
c. 1, 2, 4, 9   4
d. 3, 7, 9, 10, 10, 9   8
e. 3, 6, 9, 12, 15   9
f. 2, 4, 6, 8, 10, 12   7

Use averaging to find what number is halfway between:
g. 3 and 11   7
h. 12 and 20   16
i. 15 and 31   23
j. 24 and 42   33
k. 9 and 81   45
l. 43 and 99   71

**problem
set 18**

1. Jumbo ate two thousand, sixty-eight peanuts in the morning and three thousand, nine hundred forty in the afternoon. How many peanuts did Jumbo eat in all?   6008

2. Jimmy counted his permanent teeth. He had eleven on top and twelve on the bottom. An adult has 32 permanent teeth. How many more of Jimmy's teeth need to grow in?
9

3. Olive bought one dozen cans of spinach as a birthday present for her boyfriend. The spinach cost 53¢ per can. How much did Olive spend on spinach? $6.36

4. **Estimate** the difference of 5035 and 1987 by rounding to the nearest thousand before subtracting.   3000

*5. Find the average of 3, 7, and 8.   6

*6. What number is halfway between 27 and 81?   54

7. What number is 6 less than 2?   −4

8. $\frac{234}{6} = 39$ 9. $10,010 \div 10 =$ 1001   10. $34,180 \div 17 =$ 2010 r10

11. $364 + 9428 + 87 = 9879$ 12. $41,375 - 13,576 = 27{,}799$

13. $10 - 11 = -1$ 14. $4 \times 3 \times 2 \times 1 \times 0 = 0$

15. $125 \times 16 = 2000$ 16. $\$0.35 \times 100 = \$35$

17. The rectangle is 15 cm long and 12 cm wide. What is its perimeter?   54 cm

18. What is the sum of the first six positive even numbers?
42

19. What number is missing in this sequence? 1, 2, 4, <u> 8 </u>, 16, 32, 64

20. Compare: 500 × 1 ◯ 500 ÷ 1 =

21. (1 + 2) × 3 = (1 × 2) + ☐ 7

22. What is $\frac{1}{2}$ of 1110?   555

23. What is the place value of the 7 in 987,654,321?   millions

Use the graph to answer questions 24 and 25.

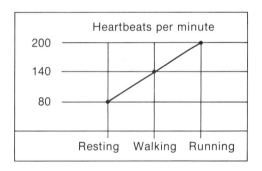

24. Running increases a resting person's heartbeat by how many beats per minute?   120

25. How many times would a person's heart beat during a 10-minute run?   2000

LESSON
19

Factors

A whole number **factor** is a whole number which divides another whole number evenly. For example, there are four whole numbers which divide 6 evenly. These four numbers are 1, 2, 3, and 6. All these numbers are factors of 6. The number 10 also has four different whole number factors. Can you think of what they are?

example 19.1 What are the whole number factors of 10?

solution The factors of 10 are all the numbers which divide 10 evenly (with no remainder). They are **1**, **2**, **5**, and **10**.

$$\mathbf{1)}\,\overline{10}\,{}^{10}, \quad \mathbf{2)}\,\overline{10}\,{}^{5}, \quad \mathbf{5)}\,\overline{10}\,{}^{2}, \quad \mathbf{10)}\,\overline{10}\,{}^{1}.$$

example 19.2 How many different whole numbers are factors of 12?

solution Twelve can be divided evenly by 1, 2, 3, 4, 6, and 12. The question asked "How many?" Counting, we find that 12 has **6** different factors.

practice Most of the time we just say **factors** instead of saying **whole number factors**. List the factors of the following numbers.
a. 14   1, 2, 7, 14
b. 15   1, 3, 5, 15
c. 16   1, 2, 4, 8, 16
d. 17   1, 17
How many different factors do each of these numbers have?
e. 18   6
f. 19   2
g. 20   6
h. 21   4

problem set 19
1. If two hundred fifty-two thousand is the dividend, and six hundred is the divisor, then what is the quotient?   420

2. Lincoln began his speech, "Fourscore and seven years ago . . . ." A score is twenty. How many years is fourscore and seven?   87

3. Overnight the temperature dropped from 4° to −3°. This was a drop of how many degrees?   7°

4. If 203 turnips are to be shared equally among seven dwarfs, how many should each receive?   29

5. What is the average of 1, 2, 4, and 9? 4

6. What is the next number in the sequence? 1, 4, 9, 16, 25, <u>36</u>

7. A regular hexagon has six sides of equal length. If each side of a hexagon is 25 mm, what is the perimeter?
150 mm

8. One centimeter equals ten millimeters. How many millimeters long is the line segment? 30 mm

*9. What are the whole number factors of 10? 1, 2, 5, 10

*10. How many different whole numbers are factors of 12? 6

11. List the whole number factors of 15. 1, 3, 5, 15

12. $250,000 \div 100 =$
2500

13. $1234 \div 60 =$
20 r34

14. $\dfrac{6 + 18 + 9}{3} = 11$

15. $\$42 + \$375 =$
$417

16. $\$3.45 \times 10 =$
$34.50

17. $1000 - 193 =$
807

18. $4 - 7 = -3$

19. $\dfrac{\square}{3} = 4$ 12

20. Compare: $123 \div 1 \bigcirc 123 \quad =$

21. Which digit is in the ten-millions' place in 135,792,468,000?
9

22. Round 123,456,789 to the nearest million. 123,000,000

23. What is $\dfrac{1}{2}$ of $11.00? $5.50

24. A square has a perimeter of 40 inches. How long is each side of the square? 10 in.

25. $(51 + 49) \times (51 - 49) = 200$

LESSON
20

Greatest Common Factor (GCF)

The factors of 8 are 1, 2, 4, and 8. The factors of 12 are 1, 2, 3, 4, 6, and 12. We see that 8 and 12 have some of the same factors. They have three factors in common. Their three common factors are 1, 2, and 4. Their **greatest common factor**—the largest factor which they both have—is 4. Greatest common factor is often abbreviated **GCF**. The letters GCF stand for **G**reatest **C**ommon **F**actor.

example 20.1 Find the greatest common factor of 12 and 18.

solution The factors of 12 are 1, 2, 3, 4, 6, and 12. The factors of 18 are 1, 2, 3, 6, 9, and 18. We see that 12 and 18 share four common factors; the largest of these four is **6**.

example 20.2 Find the GCF of 6, 9, and 15.

solution The factors of 6 are 1, 2, 3, 6. The factors of 9 are 1, 3, 9. The factors of 15 are 1, 3, 5, 15. Their GCF is **3**.
Note: The search for the greatest common factor is a search for the **largest** number which divides evenly two or more other numbers. A complete listing of the factors may be helpful but is not necessary.

practice
(See Practice
Set F in the
Appendix.)

Find the greatest common factor (GCF) of the following:
a. 10 and 15   5
b. 18 and 24   6
c. 15 and 25   5
d. 12 and 15   3

e. 18 and 27   9
f. 12, 18, and 24   6
g. 20, 30, and 40   10
h. 20, 40, and 60   20

**problem
set 20**

1. What is the difference between the product of 12 and 8 and the sum of 12 and 8?   76

2. Saturn's average distance from the sun is one billion, four hundred twenty-seven million kilometers. Write that number.   1,427,000,000

3. Which digit in 497,325,186 is in the ten-millions' place?
9

4. Ernie actually had $427,872, but when Bert asked him how much money he had, Ernie rounded his answer to the nearest thousand. How much did he say he had?
$428,000

5. The morning temperature was −3°. By afternoon it had warmed to 8°. How many degrees had the temperature risen? 11°

6. What is the average of 31, 52, and 40? 41

*7. Find the greatest common factor of 12 and 18. 6

*8. Find the GCF of 6, 9, and 15. 3

9. What is the GCF of 8 and 12? 4

10. 5432 ÷ 10 =
543 r2

11. 37,080 ÷ 12 =
3090

12. $\dfrac{28 + 42}{14} = 5$

13. 56,042 + 49,985 =
106,027

14. 14,009 − 9,670 = 4339

15. 528 + 76 = 604

16. 5 × 4 × 3 × 2 × 1 = 120

17. $6.47 × 10 = $64.70

18. Which number is missing in this sequence? _4_, 10, 16, 22, 28

19. 6 × ☐ = 90   15

20. Compare: 50 − 1 ◯ 49 + 1   <

21. How much is $\dfrac{1}{2}$ of $7.30?
$3.65

22. 365 − ☐ = 365   0

23. The first positive odd number is 1. What is the tenth positive odd number?   19

24. The perimeter of a square is 100 cm. What is the length of each side?   25 cm

25. How long is the key?

$2\frac{1}{4}$ in.

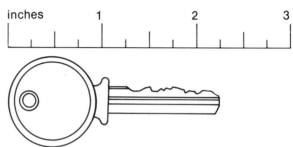

LESSON
21

Naming Fractions

Parts of whole amounts are called fractions.

A fraction may be part of 1. Here $\frac{1}{4}$ of 1 is shaded.

A fraction may be part of a group. Here $\frac{1}{4}$ of the whole group is shaded.

Common fractions are written with a top number called the **numerator** and a bottom number called the **denominator**. The numerator tells how many parts you are counting. The denominator tells how many parts are in the whole.

$$\frac{\text{Numerator}}{\text{Denominator}} \qquad \frac{1}{4}$$

example 21.1   Write the fraction three-hundredths.

solution   One hundred is the denominator, and three is the numerator.

$$\frac{3}{100}$$

example 21.2   An apple pie was cut into four equal slices. One slice was quickly eaten. What fraction of the pie was left?

solution   The fraction that remains is $\frac{3}{4}$. A picture may help. We see that when 1 part is re-moved, 3 parts remain.

example 21.3   There are 17 girls in a class of 30 students. What fraction of the class is made up of girls?

solution   There are 30 in the whole group (denomi-nator). We are counting 17 of the members (numerator).

$$\frac{17}{30}$$

practice   Write the fraction.
a. Seven-tenths $\frac{7}{10}$
b. Denominator is 6, numerator is 5. $\frac{5}{6}$
c. A team won 7 of 10 games. Fraction won? $\frac{7}{10}$
d. Another team lost 7 of 10 games. Fraction **won**? $\frac{3}{10}$

**problem
set 21**

1. What is the product of the sum of 8 and 5 and the differ-ence of 8 and 5?   39

2. In 1912 the steamship *Titanic* sank. Of the 2223 people on board, 706 were rescued. How many died?   1517

3. Tom can lift 240 pounds. That is the same as how many 16-pound bowling balls?   15

*4. An apple pie was cut into four equal slices. One slice was quickly eaten. What fraction of the pie was left? $\frac{3}{4}$

*5. There are 17 girls in a class of 30 students. What fraction of the class is made up of girls? $\frac{17}{30}$

*6. Write the fraction three-hundredths. $\frac{3}{100}$

7. What is $\dfrac{1}{2}$ of 234?   117

8. What is the place value of the 7 in 987,654,321?   millions

9. What number comes next in the sequence? 1, 4, 16, 64, ____
256

10. Compare: $64 \times 1 \bigcirc 64 \div 1$ **11.** $50 - 1 = 49 + \square$ 0
     =

12. Estimate to hundreds the sum of 396, 197, and 203.   800

13. What is the GCF of 12 and 16?   4

14. $100\overline{)4030}$ **15.** $48,840 \div 24 =$ **16.** $\dfrac{678}{6} = 113$
     40 r30   2035

17. $\$4.75 \times 10 = \$47.50$ **18.** $1000 - 87 = 913$

19. $463 + 27 + 4536 = 5026$ **20.** $3 - 10 = -7$

21. What is the average of 12, 16, and 23?   17

22. How many different whole numbers are factors of 24?   8

23. A regular octagon has eight equal-length sides. What is the perimeter of a regular octagon with sides 18 cm long?
     144 cm

24. How long is the arrow?
     $2\frac{1}{2}$ in.

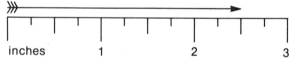
     inches   1   2   3

25. $(25 \times 25) - (24 \times 26) = 1$

LESSON
22

Fractional Part of a Group

A fraction can name part of a group. How many is $\frac{1}{2}$ of 6 jelly beans?

The group is cut into 2 equal parts by dividing. In 1 of those parts there are **3** jelly beans. Now, how many is $\frac{1}{3}$ of 6 jelly beans?

The group is cut into 3 equal parts by dividing. In 1 of those parts there are **2** jelly beans. Now, how many is $\frac{2}{3}$ of 6 jelly beans?

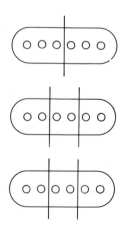

The group is cut into 3 equal parts by dividing. In each part there are 2; in two parts there are **4** jelly beans.

To find a fraction of a group, divide the group by the denominator, then multiply the answer by the numerator.

example 22.1 If $\frac{2}{3}$ of the 12 jelly beans were eaten, how many were eaten?

solution First we separate 12 into 3 parts by dividing. We find that there are 4 in each part. We want to know how many are in 2 of those parts. The answer is **8**.

example 22.2 What is $\frac{3}{4}$ of 16?

solution Divide by 4 and then multiply by 3. The answer is **12**.

practice
(See Practice
Set G in the
Appendix.)

a. What is $\frac{1}{3}$ of 21? 7

b. What is $\frac{1}{8}$ of 40? 5

c. What is $\frac{2}{3}$ of 24? 16

d. What is $\frac{3}{4}$ of 24? 18

e. What is $\frac{1}{10}$ of 80? 8

f. What is $\frac{3}{10}$ of 30? 9

g. What is $\frac{3}{5}$ of 45? 27

h. What is $\frac{5}{6}$ of 30? 25

**problem
set 22**

1. When the sum of 15 and 12 is subtracted from the product of 15 and 12, what is the difference? 153

2. There were 13 original states. There are now 50 states. What fraction of the states are the original states? $\frac{13}{50}$

3. A marathon race is 26 miles plus 385 yards. A mile is 1760 yards. Altogether, how many yards long is a marathon? 46,145 yd.

*4. If $\frac{2}{3}$ of the 12 jelly beans were eaten, how many were eaten? 8

*5. What is $\frac{3}{4}$ of 16? 12

6. What is $\frac{1}{10}$ of \$3.50?   \$0.35

7. How many spaces are between -3 and 4 on the number line?   7

8. $8 - 15 = -7$ **9.** $36 + 478 + 3409 = 3923$

10. $345 \times 67 = 23{,}115$ **11.** $\$12.45 \div 3 = \$4.15$

12. $35\overline{)1000}$ 28 r20   **13.** $\dfrac{7 + 9 + 14}{3} = 10$

14. Find the product of 36 and 124, then round the answer to the nearest hundred.   4500

15. Which digit is in the ten-millions' place in 375,426,198,000?   2

16. Find the greatest common factor of 12 and 15.   3

17. List the factors of 30.   1, 2, 3, 5, 6, 10, 15, 30

18. What is one-half of thirty thousand, four hundred?   15,200

19. Compare: $\dfrac{1}{3} + \dfrac{1}{3} \bigcirc 1$ <   **20.** $64 \div \square = 64$ 1

21. Here are the first five numbers of a sequence. What would be the **seventh** number in the sequence? 1, 4, 7, 10, 13, . . . , ___   19

22. $(3 + 3) - (3 \times 3) = -3$

23. What is the average of 27 and 43?   35

24. What is the perimeter of the rectangle?   50 cm

15 cm

10 cm

25. How long is the line segment?

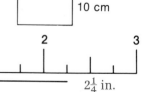

$2\frac{1}{4}$ in.

LESSON
23

Adding and
Subtracting Like Fractions

When we add, we add **like** things. When we add fractions, we add parts that have the same size.
Here is a picture of addition.

$$\frac{1}{4} \quad + \quad \frac{2}{4} \quad = \quad \frac{3}{4}$$

We see that the numerators are added while the denominator stays the same. To add fractions that have the same denominator, add the numerators only. To subtract fractions that have the same denominator, subtract the numerator only.

example 23.1 Add: $\dfrac{4}{9} + \dfrac{1}{9}$

solution $\dfrac{4}{9} + \dfrac{1}{9} = \dfrac{5}{9}$

example 23.2 Subtract: $\dfrac{7}{10} - \dfrac{4}{10}$

solution $\dfrac{7}{10} - \dfrac{4}{10} = \dfrac{3}{10}$

example 23.3 Add: $3\frac{1}{3}$
$\qquad\qquad +1\frac{1}{3}$

solution $3\frac{1}{3}$
$\qquad +1\frac{1}{3}$
$\qquad\overline{\;4\frac{2}{3}}$

practice

a. $\frac{1}{4}+\frac{1}{4}+\frac{1}{4}=\frac{3}{4}$

b. $\frac{4}{5}-\frac{1}{5}=\frac{3}{5}$

c. $1\frac{1}{5}+2\frac{2}{5}=3\frac{3}{5}$

d. $5\frac{3}{8}-1\frac{2}{8}=4\frac{1}{8}$

e. $6\frac{2}{3}-2=4\frac{2}{3}$

f. $4\frac{2}{3}-\frac{1}{3}=4\frac{1}{3}$

g. $1\frac{1}{5}+2\frac{2}{5}+3=6\frac{3}{5}$

h. $2+\frac{1}{3}=2\frac{1}{3}$

problem set 23

1. What is the difference of thirty thousand, four hundred and twenty-nine thousand, five hundred twenty-four? 876

2. How many years were there from 5 B.C. to 5 A.D.? **Note:** When spanning the years from B.C. to A.D., it must be remembered that **there was no year 0.** Therefore, the number of years is **1 less** than if the question were answered by using a number line. To solve problems like this one, **we add the years, then subtract 1.** 9 years

3. William Tell shot at the apple from 100 paces. If each pace was 36 inches, how many inches away was the apple? 3600 in.

4. There are 31 days in December. After December 25 what fraction of the month remains? $\frac{6}{31}$

5. What is $\frac{3}{5}$ of 25? 15

6. What is $\frac{1}{10}$ of 360? 36

*7. Add: $\frac{4}{9}+\frac{1}{9}=\frac{5}{9}$

*8. Subtract: $\frac{7}{10}-\frac{4}{10}=\frac{3}{10}$

*9. Add: $3\frac{1}{3}$
$+1\frac{1}{3}$
$\overline{\quad 4\frac{2}{3}}$

10. 1020
− 346
$\overline{\quad 674}$

11. 375
× 16
$\overline{\quad 6000}$

12. $\frac{375}{25}=15$

13. What is the place value of the 6 in 36,274,591? millions

14. $6-15=-9$

15. $\$0.35 \times \square = \$35.00$ 100

16. Compare: $\frac{3}{4} \bigcirc 1$ <

17. The length of a rectangle is 20 inches. The width is half the length. What is the perimeter of the rectangle? 60 in.

18. What is the sixth number in this sequence? 2, 4, 8, 16, . . . 64

19. Round 3,174 and 4,790 to the nearest thousand to estimate their sum.   8000

20. $(1234 + 766) \div (4 \times 4) = $ 125

21. What is the GCF of 24 and 32?   8

22. What is the sum of the first seven positive odd numbers?   49

Use the graph to answer questions 23, 24, and 25.

Chocolate chip cookies eaten in December	
Glen	🍪 🍪 🍪
Mark	🍪 🍪 🍪 🍪 🍪
Tony	🍪 🍪 🌙
Key:	🍪 Represents 10 cookies

23. How many more cookies were eaten by Mark than by Tony?   25

24. What was the total number eaten by all the boys?   105

25. What was the average number of cookies eaten by each boy?   35

LESSON
24

Fractions Equal to 1, Subtracting

fractions equal to 1   Each of these fractions is equal to 1.

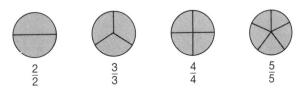

$$\frac{2}{2} \qquad \frac{3}{3} \qquad \frac{4}{4} \qquad \frac{5}{5}$$

When the numerator and denominator of a fraction are equal to each other, the fraction is equal to 1.

subtracting To solve some fraction problems we need to write 1 as a fraction. **We can change 1 to a fraction as long as the numerator and denominator are equal to each other.**

$$1 = \frac{2}{2} = \frac{3}{3} = \frac{4}{4} = \frac{5}{5} = \frac{6}{6} = \frac{7}{7} = \frac{8}{8} = \frac{n}{n}$$

example 24.1 Subtract: $1 - \frac{1}{3}$

solution A picture solution: To remove $\frac{1}{3}$ from a pie, the pie was first sliced into thirds. An arithmetic solution: Since $1 = \frac{3}{3}$ we can change the question from $1 - \frac{1}{3}$ to

$$\frac{3}{3} - \frac{1}{3} = \frac{2}{3}.$$

example 24.2 Subtract: $\begin{array}{r} 3 \\ -1\frac{1}{4} \\ \hline \end{array}$

solution To subtract we must "borrow." We borrow 1 from 3 and call the 1 we borrow $\frac{4}{4}$. Now subtract.

$$\begin{array}{r} \overset{2}{\cancel{3}} \quad \overset{1\,\searrow}{\frac{4}{4}} \\ -1 \quad \frac{1}{4} \\ \hline 1 \quad \frac{3}{4} \end{array}$$

practice
a. $1 - \frac{1}{4} = \frac{3}{4}$ **e.** $1 - \frac{3}{10} = \frac{7}{10}$ **i.** $7 - 4\frac{3}{5} = 2\frac{2}{5}$

b. $1 - \frac{2}{3} = \frac{1}{3}$ **f.** $3 - 1\frac{1}{3} = 1\frac{2}{3}$ **j.** $8 - 1\frac{1}{10} = 6\frac{9}{10}$

c. $1 - \frac{2}{5} = \frac{3}{5}$ **g.** $5 - 1\frac{3}{4} = 3\frac{1}{4}$

d. $1 - \frac{3}{7} = \frac{4}{7}$ **h.** $6 - 2\frac{1}{2} = 3\frac{1}{2}$

problem set 24

1. What is the sum of $\frac{1}{7}$, $\frac{2}{7}$, and $\frac{3}{7}$? $\frac{6}{7}$

2. Cookie ate $\frac{3}{4}$ of a dozen chocolate chip cookies in one bite. How many cookies did Cookie eat in that bite? 9

3. One mile is five thousand, two hundred eighty feet. How many feet are in $\frac{1}{10}$ of a mile?   528

*4. Subtract: $1 - \dfrac{1}{3}$ $\dfrac{2}{3}$

5. Subtract: $1 - \dfrac{3}{4}$ $\dfrac{1}{4}$

*6. Subtract: 3
$\underline{-1\frac{1}{4}}$
$1\frac{3}{4}$

7. Add: $3\frac{2}{5}$
$\underline{+1\frac{1}{5}}$
$4\frac{3}{5}$

8. Add: $\dfrac{1}{10} + \dfrac{3}{10} + \dfrac{5}{10} = \dfrac{9}{10}$

9. What number is 10 less than 3? -7

10. What number is halfway between 123 and 321?   222

11. Paul wanted to fence in a square pasture for Babe, his blue ox. Each side was to be 25 miles long. How many miles of fence did Paul need?   100 mi.

12. Round 32,987,145 to the nearest million.   33,000,000

13. What number is missing in this sequence? 1, 7, _13_, 19, 25

14. $9\overline{)1000}$ 111 r1

15. $22,422 \div 32 =$ 700 r22

16. $8 - 20 = -12$

17. $\$350.00 \div 100 = \$3.50$

18. $7 \times \boxed{} = 84$ 12

19. Compare: $\dfrac{1}{2} \bigcirc \dfrac{1}{4}$ $>$

20. What temperature is shown on the thermometer?   44°

21. $(35 \times 35) - (5 \times 5) = 1200$

22. Round 385 and 214 to hundreds to estimate their product.   80,000

23. What is the GCF of 21 and 28?   7

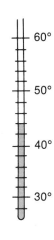

24. Of the 31 students in the class, 14 are girls. What fraction of the class is made up of boys? $\frac{17}{31}$

25. Write a fraction equal to 1 with a 4 in the denominator. $\frac{4}{4}$

LESSON
25

Multiplying and Dividing Fractions

multiplying To multiply fractions, multiply the numerators together and multiply the denominators together.

$$\frac{1}{2} \times \frac{3}{4} = \frac{3}{8}$$

of When we use the word **of** with a fraction we mean to multiply.

$$\frac{3}{4} \text{ of } \frac{4}{5} \quad \text{means} \quad \frac{3}{4} \times \frac{4}{5}$$

dividing We divide by a fraction by changing the problem to a multiplication problem. To do this we invert the divisor and multiply. The divisor is the fraction after the division symbol ($\div$). We invert the divisor by writing the top term on the bottom and the bottom term on the top.

Change the division problem $\quad \frac{1}{2} \div \frac{2}{3} =$

to a multiplication problem. $\quad \frac{1}{2} \times \frac{3}{2} = \frac{3}{4}$

example 25.1 $\quad \frac{3}{8} \times \frac{1}{2} =$

solution When we multiply fractions, we multiply the numerators and we multiply the denominators.

$$\frac{3}{8} \times \frac{1}{2} = \frac{3}{16}$$

example 25.2 $\dfrac{2}{3} \div \dfrac{3}{4} =$

solution   To divide fractions, we invert the second fraction and multiply.

$$\dfrac{2}{3} \div \dfrac{3}{4} =$$

$$\dfrac{2}{3} \times \dfrac{4}{3} = \dfrac{8}{9}$$

practice a. $\dfrac{3}{4} \times \dfrac{1}{2} = \dfrac{3}{8}$ e. $\dfrac{1}{2} \times \dfrac{1}{2} \times \dfrac{1}{2} = \dfrac{1}{8}$ i. $\dfrac{2}{5} \div \dfrac{3}{2} = \dfrac{4}{15}$

  b. $\dfrac{1}{2}$ of $\dfrac{3}{4} = \dfrac{3}{8}$ f. $\dfrac{1}{2} \div \dfrac{2}{3} = \dfrac{3}{4}$ j. $\dfrac{2}{7} \div \dfrac{3}{4} = \dfrac{8}{21}$

  c. $\dfrac{1}{3} \times \dfrac{1}{2} = \dfrac{1}{6}$ g. $\dfrac{1}{3} \div \dfrac{1}{2} = \dfrac{2}{3}$

  d. $\dfrac{1}{4}$ of $\dfrac{3}{4} = \dfrac{3}{16}$ h. $\dfrac{1}{4} \div \dfrac{1}{3} = \dfrac{3}{4}$

**problem
set 25**

1. What is the product of the sum of 55 and 45 and the difference of 55 and 45?   1000

2. Potatoes are three-fourths water. If a sack of potatoes weighs 20 pounds, how many pounds are water?   15 lb.

3. Frankie found three hundred six fleas on his dog. He caught two hundred forty-nine of them. How many fleas got away?
57

4. What number is halfway between 8 and 9? $8\frac{1}{2}$

5. What number is 15 less than 3? -12

6. Round 1,234,567 to the nearest ten thousand.   1,230,000

7. Five of the dozen donuts were chocolate. What fraction were not chocolate? $\frac{7}{12}$

8. What is the denominator of $\dfrac{23}{24}$?   24

9. What is $\dfrac{1}{5}$ of 65?   13   10. What is $\dfrac{2}{3}$ of 15?   10

11. $\frac{1}{10} + \frac{1}{10} + \frac{1}{10} =$ **12.** $3\frac{1}{3} + 4 = 7\frac{1}{3}$ **13.** $1 - \frac{3}{5} = \frac{2}{5}$
$\frac{3}{10}$

14. $3 - 1\frac{1}{4} = 1\frac{3}{4}$ ***15.** $\frac{3}{8} \times \frac{1}{2} = \frac{3}{16}$ ***16.** $\frac{2}{3} \div \frac{3}{4} = \frac{8}{9}$

17. What is $\frac{1}{2}$ of $\frac{1}{4}$? $\frac{1}{8}$

18. $45.60 \div 10 =$ **19.** $52 \overline{)2100}$ **20.** $\frac{432}{18} = 24$
4.56 40 r20

21. If a 36-inch-long string is made into the shape of a square, how long will each side be? 9 in.

22. Compare: $1 \times 2 \times 3 \times 4 \bigcirc 2 \times 3 \times 4 =$

23. $(55 + 45) \times (55 - 45) = 1000$

24. Three hundred seventy-five is equal to two hundred plus what number? 175

25. How many quarters equal twenty-five dollars? 100

LESSON 26   **Reciprocals**

Whole numbers tell how many wholes. We can write a 1 under any whole number without changing it.

$$4 \text{ is the same as } \frac{4}{1} \qquad \frac{5}{1} \text{ is the same as } 5$$

A number with the numerator and denominator reversed is the **reciprocal** of the number.

$$\frac{2}{3} \text{ is the reciprocal of } \frac{3}{2} \qquad \frac{3}{2} \text{ is the reciprocal of } \frac{2}{3}$$

$$\frac{1}{5} \text{ is the reciprocal of } 5 \qquad 5 \text{ is the reciprocal of } \frac{1}{5}$$

The product of a number and the reciprocal of the same number is always 1.

$$\frac{2}{3} \times \frac{3}{2} = 1 \qquad \frac{4}{5} \times \frac{5}{4} = 1 \qquad \frac{11}{7} \times \frac{7}{11} = 1 \qquad 4 \times \frac{1}{4} = 1$$

In the last lesson, we learned to divide fractions by inverting and multiplying. When we invert, we get the reciprocal. Thus, we divide by multiplying by the reciprocal.

example 26.1 Simplify: $\dfrac{2}{5} \div \dfrac{3}{7}$

solution To divide fractions, we multiply by the reciprocal of the divisor. $\qquad \dfrac{2}{5} \times \dfrac{7}{3} = \dfrac{14}{15}$

practice **a.** $\dfrac{2}{3} \times \square = \dfrac{6}{6}$ $\dfrac{3}{2}$

b. $\dfrac{3}{4} \times \square = \dfrac{12}{12}$ $\dfrac{4}{3}$

c. $\dfrac{5}{3} \times \square = 1$ $\dfrac{3}{5}$

d. $\dfrac{6}{1} \times \square = 1$ $\dfrac{1}{6}$

e. What is the reciprocal of $\dfrac{2}{5}$? $\dfrac{5}{2}$

f. What is the reciprocal of $\dfrac{3}{1}$? $\dfrac{1}{3}$

g. Instead of dividing $\frac{1}{2}$ by $\frac{3}{4}$, you could multiply $\frac{1}{2}$ by what? $\frac{4}{3}$

problem set 26

1. What is the difference between the sum of $\frac{1}{4}$ and $\frac{1}{4}$ and the product of $\frac{1}{2}$ and $\frac{1}{2}$? $\frac{1}{4}$

2. In three tries Carlos punted the football 35 yards, 30 yards, and 37 yards. What was the average distance of his punts?
   34 yd.

3. The earth's average distance from the sun is one hundred forty-nine million, six hundred thousand kilometers. Write that number. 149,600,000

4. What is the perimeter of the rectangle? 1 in. $\frac{3}{8}$ in.

5. How many years were there from 50 B.C. to 50 A.D.? 99 years $\frac{1}{8}$ in.

6. If $\dfrac{1}{3}$ of the jelly beans were eaten, what fraction is left?
   $\frac{2}{3}$

7. Compare: $\dfrac{1}{2}$ of 12 $\bigcirc$ $\dfrac{1}{3}$ of 12 $>$

8. $\dfrac{1}{2} \times \dfrac{1}{2} \times \dfrac{1}{2} = \dfrac{1}{8}$ **9.** $\dfrac{2}{3} \div \dfrac{3}{2} = \dfrac{4}{9}$ **10.** $\dfrac{2}{3} \times \dfrac{3}{2} = 1$

11. $1 + \dfrac{2}{3} = 1\dfrac{2}{3}$ **12.** $4 - 1\dfrac{3}{5} = 2\dfrac{2}{5}$ **13.** $\dfrac{2}{5} \div \dfrac{3}{7} = \dfrac{14}{15}$

14. What is the reciprocal of $\dfrac{3}{4}$? $\dfrac{4}{3}$

15. To divide $\dfrac{1}{2}$ by $\dfrac{2}{3}$ we multiply $\dfrac{1}{2}$ by what number? $\dfrac{3}{2}$

16. $\dfrac{4}{5} \times \Box = 1$ $\dfrac{5}{4}$

17. What number comes next? 81, 64, 49, 36, <u>25</u>

18. Cheryl bought 10 pens for $0.25 each. How much did she pay in all? $2.50

19. What is the greatest common factor (GCF) of 24 and 32? 8

20. $(30 \times 40) \div 60 = 20$ **21.** What is $\dfrac{1}{100}$ of 100?   1

22. Estimate the sum of 3142, 6328, and 4743 to the nearest thousand.   14,000

23. Two-thirds of the students liked hamburgers more than hot dogs. If 60 students were asked, how many liked hamburgers better?   40

24. $\dfrac{144}{12} = 12$

25. How long is the line segment?

$1\dfrac{3}{4}$ in.

LESSON 27

Converting Improper Fractions to Whole Numbers or Mixed Numbers

When the numerator (top) of a fraction is the same as the denominator (bottom) of the fraction, the fraction equals 1. When the numerator (top) of a fraction is smaller than the denominator (bottom) of the fraction, the fraction is called a **proper fraction**. When the numerator (top) of the fraction is equal to or greater than the denominator (bottom) of the fraction, the fraction is called an **improper fraction**.

Each of these fractions is an improper fraction.

$$\frac{2}{2} \qquad \frac{23}{10} \qquad \frac{1001}{1000}$$

Recall that the bar used for a fraction line also means to divide. We can change an improper fraction to a mixed number by dividing.

Improper fractions are simplified by dividing and writing the result as a whole number or as a whole number plus a fraction.

example 27.1 Write $\dfrac{14}{5}$ as a mixed number.

solution We divide and write the result as a mixed number.

$$
\begin{array}{r}
2 \\
5\overline{)14} \\
\underline{10} \\
4 \quad \longleftarrow \text{Remainder}
\end{array}
$$

$$\frac{14}{5} = 2\frac{4}{5} \quad \begin{array}{l} \longleftarrow \text{Remainder} \\ \longleftarrow \text{Same denominator} \end{array}$$

example 27.2 Add and simplify: $\dfrac{2}{5} + \dfrac{7}{5}$

solution $\dfrac{2}{5} + \dfrac{7}{5} = \dfrac{9}{5}$

$$
\begin{array}{r}
1 \\
5\overline{)9} \\
\underline{5} \\
4 \quad \longleftarrow \text{Remainder}
\end{array}
$$

$$\frac{9}{5} = 1\frac{4}{5} \quad \begin{array}{l} \longleftarrow \text{Remainder} \\ \longleftarrow \text{Same} \\ \text{denominator} \end{array}$$

practice Convert the improper fractions to whole or mixed numbers.

a. $\dfrac{6}{2} = 3$

b. $\dfrac{7}{3} = 2\dfrac{1}{3}$

c. $\dfrac{15}{4} = 3\dfrac{3}{4}$

d. $\dfrac{5}{5} = 1$

e. $\dfrac{3}{5} + \dfrac{4}{5} = 1\dfrac{2}{5}$

f. $\dfrac{2}{3} + \dfrac{2}{3} + \dfrac{2}{3} = 2$

g. $\dfrac{5}{2} \times \dfrac{4}{5} = 2$

h. $\dfrac{1}{2} \div \dfrac{1}{3} = 1\dfrac{1}{2}$

problem   1. What is the difference between the sum of $\frac{1}{2}$ and $\frac{1}{2}$ and the
set 27   product of $\frac{1}{2}$ and $\frac{1}{2}$? $\frac{3}{4}$

2. The young elephant was 36 months old. How many years old was the elephant?   3 years

3. Gwen brought $2\frac{1}{2}$ dozen cupcakes for the party. That was enough for how many children to have one cupcake each?
   30

4. There are 100 centimeters in a meter. There are 1000 meters in a kilometer. How many centimeters are in a kilometer?
   100,000 cm

5. What is the perimeter of the equilateral triangle?   2 in.

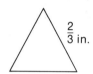

6. Compare: $\dfrac{1}{2} + \dfrac{1}{2} \bigcirc \dfrac{1}{2} \times \dfrac{1}{2}$ 7. $\dfrac{2}{5} \times \square = 1$ $\dfrac{5}{2}$
   >

8. What is the reciprocal of $\dfrac{5}{6}$? $\dfrac{6}{5}$

9. What is the denominator of $\dfrac{3}{10}$?   10

10. What is the greatest common factor of 15 and 25?   5

11. What is the **seventh** number in this sequence? 8, 16, 24, 32, 40, . . .   56

*12. Write $\dfrac{14}{5}$ as a mixed number. $2\dfrac{4}{5}$

*13. Add and simplify: $\dfrac{2}{5} + \dfrac{7}{5}$ $1\dfrac{4}{5}$

14. $\dfrac{2}{3} + \square = 1$ $\dfrac{1}{3}$

15. What is the largest number which divides evenly into both 12 and 18?   6

16. $5 - 1\dfrac{3}{4} = 3\dfrac{1}{4}$ **17.** $\dfrac{2}{3} \times \dfrac{2}{3} = \dfrac{4}{9}$ **18.** $\dfrac{1}{3} \div \dfrac{3}{4} = \dfrac{4}{9}$

19. What number is 25 less than 10? −15

20. $(123 + 123 + 123) - (123 + 123) =$ 123

21. Estimate the difference of 5063 and 3987 to the nearest thousand. 1000

22. How many fifths are in 1? 5

23. Find the average of 85, 85, 90, and 100. 90

24. What is $\frac{3}{5}$ of 30? 18

25. How many millimeters long is the line segment?
26 mm

LESSON
28

Converting Mixed Numbers

The fraction line means to divide. When the numerator (top) of the fraction is larger than the denominator (bottom) of the fraction, the fraction is an improper fraction. In Lesson 27 we found that some improper fractions represent whole numbers while other improper fractions represent mixed numbers.

(a) $\dfrac{16}{8} \longrightarrow 8\overline{\smash{)}\,16}$ so $\dfrac{16}{8} = 2$
$\qquad\qquad\quad \underline{16}$
$\qquad\qquad\quad\ \ 0$

(b) $\dfrac{5}{3} \longrightarrow 3\overline{\smash{)}\,5}$ so $\dfrac{5}{3} = 1\dfrac{2}{3}$
$\qquad\qquad\quad \underline{3}$
$\qquad\qquad\quad 2$

If we encounter a whole number and an improper fraction, the improper fraction should be simplified. Then it is added to the whole number. The numbers shown here can be simplified.

(c) $4\dfrac{3}{3}$ (d) $5\dfrac{4}{3}$

We simplify the improper fractions and add:

(c) $4\dfrac{3}{3} = 4 + 1 = 5$ (d) $5\dfrac{4}{3} = 5 + 1\dfrac{1}{3} = 6\dfrac{1}{3}$

example 28.1   Simplify: $6\dfrac{5}{3}$

solution   Convert $\frac{5}{3}$ to $1\frac{2}{3}$ and add to 6:

$$6\dfrac{5}{3} = 6 + 1\dfrac{2}{3} = 7\dfrac{2}{3}$$

example 28.2   Simplify: $1\dfrac{2}{3} + 1\dfrac{2}{3}$

solution   Add to get $2\frac{4}{3}$, then convert the improper fraction.

$$1\dfrac{2}{3} + 1\dfrac{2}{3} = 2\dfrac{4}{3} = 2 + 1\dfrac{1}{3} = 3\dfrac{1}{3}$$

practice
(See Practice
Set H in the
Appendix.)

a. $5\dfrac{2}{2} = 6$

b. $6\dfrac{4}{2} = 8$

c. $7\dfrac{3}{2} = 8\dfrac{1}{2}$

d. $8\dfrac{5}{2} = 10\dfrac{1}{2}$

e. $9\dfrac{5}{3} = 10\dfrac{2}{3}$

f. $1\dfrac{2}{3} + 2\dfrac{1}{3} = 4$

g. $2\dfrac{2}{3} + 1\dfrac{2}{3} = 4\dfrac{1}{3}$

h. $1\dfrac{1}{3} + 1\dfrac{1}{3} + 1\dfrac{1}{3} = 4$

i. $3\dfrac{4}{5} + 1\dfrac{3}{5} = 5\dfrac{2}{5}$

j. $2\dfrac{2}{3} + 2\dfrac{2}{3} + 2\dfrac{2}{3} =$
   8

problem
set 28

1. What is the sum of $\dfrac{1}{3}$ and $\dfrac{2}{3}$ and $\dfrac{3}{3}$?   2

2. Two-fifths of Robin Hood's one hundred forty men rode with Little John to the castle. How many men went with Little John?   56

3. Seven hundred sixty-eight peanuts are to be shared equally by the thirty-two children at the party. How many should each receive?   24

4. Augustus, the first Roman Emperor, was born in 63 B.C. and died in 14 A.D. How old was he when he died?
  76 years

*5. Simplify: $6\dfrac{5}{3}$ $7\frac{2}{3}$ *6. $1\dfrac{2}{3} + 1\dfrac{2}{3} = 3\dfrac{1}{3}$ 7. $3 + 4\dfrac{2}{3} = 7\dfrac{2}{3}$

8. $3 - 1\frac{1}{5} = 1\frac{4}{5}$ 9. $\frac{1}{2} \times \frac{1}{3} \times \frac{1}{4} = \frac{1}{24}$ 10. $\frac{8}{10} + \frac{3}{10} = 1\frac{1}{10}$

11. $\frac{2}{3}$ of 24 = 16   12. $\frac{1}{5} + \frac{2}{5} + \frac{3}{5} = 1\frac{1}{5}$

13. What number is 24 less than 8? -16 14. $\frac{1}{4} \times \square = 1$ $\frac{4}{1}$ (or 4)

15. Compare: 25 × 25 $\bigcirc$ 26 × 24 >

16. On the last four papers Christie had 22 right, 20 right, 23 right, and 23 right. She averaged how many right on each paper? 22

17. A 36-inch-long string is formed into the shape of an equilateral triangle. How long is each side? 12 in.

18. What is the greatest common factor (GCF) of 24, 36, and 60? 12

19. 10,010 − 9,909 = 101   20. (100 × 100) − (100 × 99) = 100

21. If $\frac{1}{10}$ of the class was absent, what fraction of the class was present? $\frac{9}{10}$

22. 423 − $\square$ = 297   126   23. 5096 ÷ 10 = 509 r6

24. How many eggs are $\frac{3}{4}$ of two dozen? 18

25. How long is the line segment? $3\frac{1}{4}$ in.

LESSON
29

Reducing Equivalent Fractions

Different fractions which name the same number are **equivalent fractions**. Equivalent fractions are equal to each other. We can see equivalent fractions with pictures. We see that $\frac{1}{2}, \frac{2}{4}, \frac{3}{6}$, and $\frac{4}{8}$ are equivalent fractions.

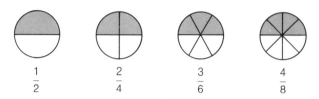

$$\frac{1}{2} \qquad \frac{2}{4} \qquad \frac{3}{6} \qquad \frac{4}{8}$$

Each of these fractions equals $\frac{1}{2}$. We can **reduce** the other fractions to $\frac{1}{2}$. When we do this, we say we are writing the fractions in **lowest terms**. To reduce a fraction to lowest terms, we divide the numerator and the denominator by the largest whole number that will divide both evenly.

$$\frac{2 \div 2}{4 \div 2} = \frac{1}{2} \qquad \frac{3 \div 3}{6 \div 3} = \frac{1}{2} \qquad \frac{4 \div 4}{8 \div 4} = \frac{1}{2}$$

example 29.1 Reduce: $\dfrac{6}{8}$

solution Both 6 and 8 can be divided by 2, so we divide 6 by 2 and divide 8 by 2. $\dfrac{6 \div 2}{8 \div 2} = \dfrac{3}{4}$

example 29.2 Add and reduce: $\dfrac{1}{8} + \dfrac{3}{8}$

solution We add. Then we divide 4 by 4 and divide 8 by 4.

$$\frac{1}{8} + \frac{3}{8} = \frac{4}{8} \qquad \text{then} \qquad \frac{4 \div 4}{8 \div 4} = \frac{1}{2}$$

practice
(See Practice Sets I, J, and K in the Appendix.)

Reduce to lowest terms.

a. $\dfrac{5}{10} \quad \dfrac{1}{2}$

b. $\dfrac{6}{9} \quad \dfrac{2}{3}$

c. $\dfrac{8}{12} \quad \dfrac{2}{3}$

d. $\dfrac{6}{10} \quad \dfrac{3}{5}$

e. $\dfrac{1}{4} + \dfrac{1}{4} = \dfrac{1}{2}$

f. $\dfrac{5}{6} - \dfrac{1}{6} = \dfrac{2}{3}$

g. $\dfrac{1}{2} \times \dfrac{2}{3} = \dfrac{1}{3}$

h. $\dfrac{1}{4} \div \dfrac{1}{2} = \dfrac{1}{2}$

problem set 29

1. If the sum of $\frac{1}{3}$ and $\frac{1}{3}$ is divided by the product of $\frac{1}{2}$ and $\frac{1}{2}$, what is the quotient? $2\frac{2}{3}$

2. The African elephant can weigh eight tons. A ton is two thousand pounds. How many pounds can an African elephant weigh? 16,000 lb.

3. Sixteen jelly beans weigh one ounce. Oliver ate one pound of jelly beans. How many jelly beans did Oliver eat? (1 pound = 16 ounces.) 256

*4. Reduce: $\dfrac{6}{8}$ $\dfrac{3}{4}$

*5. Add and reduce: $\dfrac{1}{8} + \dfrac{3}{8} =$ $\frac{1}{2}$

6. $\dfrac{3}{4} - \dfrac{1}{4} = \dfrac{1}{2}$

7. $\dfrac{1}{2} \times \dfrac{2}{3} = \dfrac{1}{3}$

8. $\dfrac{1}{2} \div \dfrac{1}{4} = 2$

9. What is $\dfrac{1}{10}$ of 4000? 400

10. Write the next three numbers in the sequence. 1, 4, 7, 10, ___, ___, ___ 13, 16, 19

11. When five months have passed, what fraction of the year remains? $\frac{7}{12}$

12. $3.60 \times 100 =$ $360

13. $50,000 \div 100 =$ 500

14. $\dfrac{15}{4} = 3\dfrac{3}{4}$

15. The temperature rose from $-8°$ to $15°$. This was a rise of how many degrees? 23°

16. $37 + 496 + 2684 =$ 3217

17. $1000 - 143 =$ 857

18. $7 \times 11 \times 13 =$ 1001

19. $\boxed{} \times 24 = 480$ 20

20. Round 4963 to thousands and divide by 39 rounded to tens to **estimate** the quotient. 125

21. Compare: $\dfrac{375}{375} \bigcirc 1$ $=$

22. The perimeter of the rectangle is 60 mm. Its width is 10 mm. What is its length? 20 mm

?
$\boxed{}$ 10 mm

23. $12 - 40 = -28$

24. $\left(\dfrac{1}{2} \times \dfrac{1}{2}\right) - \dfrac{1}{4} = 0$

25. How long is the line segment?

46 mm

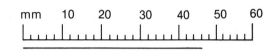

mm   10   20   30   40   50   60

LESSON
30

Least Common Multiple

We find **multiples** of a number by multiplying the number by 1, 2, 3, 4, 5, 6 and so on.

The first six multiples of 2 are 2, 4, 6, 8, 10, 12.
The first six multiples of 3 are 3, 6, 9, 12, 15, 18.
The first six multiples of 4 are 4, 8, 12, 16, 20, 24.
The first six multiples of 5 are 5, 10, 15, 20, 25, 30.

Common multiples are numbers which are multiples of more than one number. In the list of the multiples of 2 and 3, we see that 6 and 12 are common multiples of 2 and 3. Since 6 is less than 12 it is called the **least common multiple** of 2 and 3. The letters LCM are often used to stand for **Least Common Multiple**.

example 30.1 What are the first four multiples of 8?

solution Multiplying 8 by 1, 2, 3, and 4 gives the first four multiples: **8, 16, 24, 32**.

example 30.2 What is the least common multiple of 2 and 4?

solution Looking at the multiples of 2 and 4 we see 4, 8, and 12 in both lists. The least is **4**.

example 30.3 What is the LCM of 3 and 4?

solution The smallest number which is a multiple of both 3 and 4 is **12**.

practice
(See Practice
Set L in the
Appendix.)

a. What does LCM stand for?   least common multiple

b. What are the first four multiples of 9?   9, 18, 27, 36

c. What is the LCM of 2 and 5?   10

d. What is the LCM of 3 and 5?   15

e. What is the LCM of 2, 3, and 4?   12

**problem
set 30**

1. If the fourth multiple of 3 is subtracted from the third multiple of 4, what is the difference?   0

2. About $\frac{2}{3}$ of a person's body weight is water. Albert weighs 117 pounds. How many pounds of Albert's weight is water?   78 lb.

3. Cynthia ate 42 pieces of popcorn in the first 15 minutes of a movie. If she kept eating at the same rate, how many pieces of popcorn did she eat in the 2-hour movie?   336

*4. What are the first four multiples of 8?   8, 16, 24, 32

*5. What is the least common multiple of 2 and 4?   4

*6. What is the LCM of 3 and 4?   12

7. $\frac{2}{5} + \frac{2}{5} + \frac{2}{5} = 1\frac{1}{5}$ 8. $1 - \frac{1}{10} = \frac{9}{10}$ 9. $3 - 1\frac{1}{10} = 1\frac{9}{10}$

10. $\frac{3}{4} \times \frac{4}{3} = 1$ 11. $\frac{2}{3} \div \frac{3}{4} = \frac{8}{9}$ 12. $5\frac{1}{2} - 3\frac{1}{2} = 2$

13. The number 24 has how many different whole number factors?   8

14. $3 + 24 + 6.50 = $ \$33.50   15. $5 - 1.50 = $ \$3.50

16. Estimate the product: $596 \times 405 = $ 240,000

17. Find the difference of one billion and nine hundred eight million, fifty-three thousand.   91,947,000

18. Compare: $\frac{2}{3} \times \frac{2}{2} \bigcirc \frac{2}{3} \times 1$ 19. $500,000 \div 100 = $ 5000
    $=$

20. 35) 8540   244

21. $\dfrac{100}{7}$ $14\dfrac{2}{7}$

22. Four is what fraction of twelve? (Reduce.) $\frac{1}{3}$

23. What is the average of 375, 632, and 571?   526

24. A regular hexagon has six equal sides. If a regular hexagon is made from a 36-inch-long string, how long will each side be?   6 in.

25. What is the product of a number and its reciprocal?   1

LESSON 31   Area

We have measured the distance around shapes. The distance around a shape is called its **perimeter**. We measure perimeters with units of length like centimeters or inches.

We can also measure how much surface is enclosed by the sides of a shape. When we measure the "insides" of a flat shape we are measuring its **area**. The area of a shape is the number of squares (such as floor tiles) of a certain size that will cover the shape. The size of each "floor tile" is given in square units. We abbreviate the word "square" by writing sq.

 1 square centimeter (sq. cm)

We find the number of square units that will cover a rectangle by multiplying the length by the width.

example 31.1   How many square floor tiles 1 centimeter on a side would be needed to cover this rectangle?

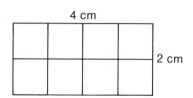

solution   If we count, we get 8 square floor tiles. If we multiply, we get

$$4 \text{ cm} \times 2 \text{ cm} = 8 \text{ sq. cm}$$

so **8** square floor tiles are needed.

example 31.2   What is the area of the square?

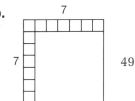

solution   All sides of a square are the same length. Thus, both the length and the width are 10 ft. We find the area by multiplying.

$$10 \text{ ft.} \times 10 \text{ ft.} = \textbf{100 sq. ft.}$$

Many of the area diagrams in this book will not be drawn to scale. The square in this problem is not drawn to scale as the sides shown here are really not 10 ft. long.

practice   Find the number of square units needed to cover the area of these shapes.

a.
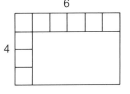
24

b.
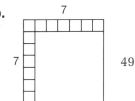
49

c.
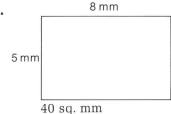
40 sq. mm

d.
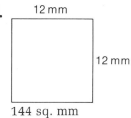
144 sq. mm

problem set 31

1. When the third multiple of 4 is divided by the fourth multiple of 3, what is the quotient?   1

2. What time is three and one-half hours after 6:50 A.M.?
10:20 A.M.

3. The distance the earth travels around the sun each year is about five hundred eighty million miles. Write that number.   580,000,000

4. Convert $\dfrac{10}{3}$ to a mixed number. $3\dfrac{1}{3}$

*5. How many square floor tiles 1 centimeter on a side would be needed to cover this rectangle?   8 sq. cm

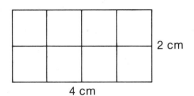

*6. How many square floor tiles 1 foot on a side would be needed to cover this square?   100 sq. ft.

7. What is the area of a rectangle 12 inches long and 8 inches wide?   96 sq. in.

8. What is the next number in the sequence? 1, 4, 9, 16, 25, 36, __49__

9. What is $\dfrac{2}{3}$ of 24?   16   **10.** $24 + \square = 42$ 18

Write each answer in simplest form.

11. $\dfrac{1}{8} + \dfrac{1}{8} = \dfrac{1}{4}$ **12.** $\dfrac{5}{6} - \dfrac{1}{6} = \dfrac{2}{3}$ **13.** $\dfrac{2}{3} \times \dfrac{1}{2} = \dfrac{1}{3}$ **14.** $\dfrac{1}{4} \div \dfrac{3}{4} = \dfrac{1}{3}$

15. Estimate the product of 387 and 514.   200,000

16. $2000 \div 10 =$ **17.** $47 \times 63 =$ **18.** $4623 \div 22 =$
200   2961   210 r3

19. $(1000 - 987) + (234 - 56) =$ 191

20. $56 - 65 =$ -9

21. Which is closest to 100? (a) 90 (b) 89 (c) 111 (d) 109
(d) 109

22. Which digit is in the ten-millions' place in 987,654,321?   8

23. The first three positive odd numbers are 1, 3, and 5. What is the sum of the first five positive odd numbers?   25

24. Three of the nine players play outfield. What fraction of the players play outfield? (Reduce.) $\frac{1}{3}$

25. How long is the line segment? $2\frac{3}{4}$ in.

inches   1   2   3   4

LESSON

32

Decimal Place Value

In our number system the **places** that digits occupy have a value, called **place value**. The value of each place is one-tenth the value of the place to its left.

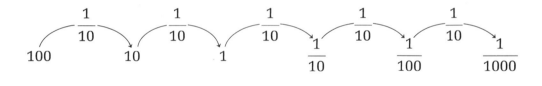

The decimal point shows where the whole number part of a number ends and where the fraction part begins. The place to the **left** of the point is always the **ones'** place. The place to the **right** of the point is always the **tenths'** place.

example 32.1   What is the place value of the 7 in 45.67?

solution   Count place value from the **decimal point**, not from the end of the number. The value of the place two places to the right of the decimal point is $\frac{1}{100}$, so the place is the **hundredths'** place.

example 32.2   Which digit is in the ones' place in 123.456?

solution   The ones' place is always the place just to the left of the decimal point. The digit in the ones' place is **3**.

practice   **a.** What is the place value of the 5 in 12.345?   thousandths
b. Which digit in 5.4321 is in the tenths' place?   4
c. In 0.0123, what is the digit in the thousandths' place?   2
d. What is the value of the place held by zero in 50.375?   ones
e. How many ones does it take to equal 10?   10
f. How many tenths does it take to equal 1.0?   10
g. How many tens does it take to equal 100?   10

**problem
set 32**

1. When the sum of 24 and 7 is multiplied by the difference of 18 and 6, what is the product?   372

2. Davy Crockett was born in Tennessee in 1786 and died at the Alamo in 1836. How many years did he live?   50 years

3. A 16-ounce box of a certain cereal costs $2.24. What is the cost per ounce of this cereal? $0.14

*4. What is the place value of the 7 in 45.67?   hundredths

*5. Which digit is in the ones' place in 123.456?   3

6. Which digit in 12.34 is in the tenths' place?   3

7. How many **square centimeters** would be needed to cover the area of the rectangle?   40 sq. cm

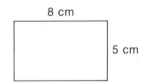
8 cm

5 cm

8. How many **centimeters** are needed to reach around the same rectangle?   26 cm

9. What is the **eighth** number in this sequence? 1, 3, 5, 7, . . .
15

10. Which digit in 1234 has the same place value as the 6 in 567?   3

11. Which number is closest to 1000? (a) 990 (b) 909 (c) 1009
(d) 1090   (c) 1009

12. Find the average of 623, 494, and 380.   499

13. $0.05 × 100 =   **14.** 8 × ☐ = 240   **15.** What is $\frac{3}{4}$ of 24?

$5   30   18

Write each answer in simplest form.

16. $\frac{3}{5} + \frac{3}{5} = 1\frac{1}{5}$ **17.** $\frac{3}{4} - \frac{1}{4} = \frac{1}{2}$ **18.** $\frac{3}{4} \times \frac{1}{3} = \frac{1}{4}$ **19.** $\frac{3}{4} \div \frac{1}{3} = 2\frac{1}{4}$

20. What number is halfway between 26 and 62?   44

21. 12 − 53 = −41   **22.** $1\frac{2}{3} - \frac{4}{4} = \frac{2}{3}$

23. (45 × 45) − (200 × 10) =   **24.** What is the least common
25   multiple of 4 and 6?   12

25. How long is the line segment? $3\frac{1}{4}$ in.

LESSON 33

Expanded Notation

1000	100	10	1	$\frac{1}{10}$	$\frac{1}{100}$	$\frac{1}{1000}$
thousands	hundreds	tens	ones	tenths	hundredths	thousandths

One way of naming numbers is to say how many of each place value a number has. For example, 234 could be named: 2 hundreds and 3 tens and 4 ones. We could write the same thing this way: (2 × 100) + (3 × 10) + (4 × 1). This form of writing a number is called **expanded notation**.

There are many ways that we can use digits to write a particular number. We say that each of the ways is a different **numeral** for the same number.

example 33.1   Write 3040 in expanded notation.

solution   There are 3 thousands and 4 tens, so write **(3 × 1000) + (4 × 10)**. Note: We do not write place values held by zero.

example 33.2   Write (6 × 100) + (2 × 1) in standard notation.

solution   Standard notation is our usual way of writing numbers. We write a 6 in the hundreds' place and a 2 in the ones' place. There are no tens, so in the tens' place we write 0.

100	10	1
6	**0**	**2**

Place value

The number (numeral)

Thus (6 × 100) + (2 × 1) equals 602.

example 33.3   Write 4.6 in expanded notation.

solution   **(4 × 1) + (6 × $\frac{1}{10}$)**

practice
(See Practice
Set M in the
Appendix.)

Write these numbers in expanded notation.

a. 501   (5 × 100) + (1 × 1)
b. 5010   (5 × 1000) + (1 × 10)
c. 630   (6 × 100) + (3 × 10)
d. 4.7   (4 × 1) + (7 × $\frac{1}{10}$)
e. 0.96   (9 × $\frac{1}{10}$) + (6 × $\frac{1}{100}$)
f. 30.25
(3 × 10) + (2 × $\frac{1}{10}$) + (5 × $\frac{1}{100}$)

Write these numbers in standard notation.

g. (5 × 100) + (3 × 10)   530
h. (6 × 1000) + (4 × 10)   6040
i. (2 × 1000) + (1 × 1)   2001
j. (6 × 1) + (4 × $\frac{1}{10}$)   6.4
k. (5 × $\frac{1}{10}$) + (7 × $\frac{1}{100}$)   0.57
l. (2 × 1) + (3 × $\frac{1}{10}$) + (4 × $\frac{1}{100}$)
2.34

problem
set 33

1. When the product of 10 and 15 is divided by the sum of 10 and 15, what is the quotient?   6

2. The Nile River is 6,651 kilometers long. The Mississippi is 5,986 kilometers long. How much longer is the Nile?
665 km

3. Some astronomers think the universe may be fifteen billion years old. Write that number.   15,000,000,000

*4. Write 3040 in expanded notation.   (3 × 1000) + (4 × 10)

*5. Write (6 × 100) + (2 × 1) in standard notation.   602

6. Write (1 × 10) + (2 × $\frac{1}{10}$) in standard notation.   10.2

7. Which digit in 452.367 is in the hundredths' place?   6

8. What is the perimeter of the rectangle?
40 in.

12 in.

8 in.

9. How many square tiles 1 inch on a side would be needed to cover the rectangle?   96 sq. in.

10. Which of these is least? (a) 36,428 (b) 29,899 (c) 111,100
(b) 29,899

11. Estimate the difference of 4968 and 2099.   3000

12. 4300 × 100 =
430,000

13. ☐ − 24 = 23
47

14. $\frac{3}{5}$ of 20 = 12

Write each answer in simplest form.

15. $\frac{4}{5} + \frac{4}{5} = 1\frac{3}{5}$ **16.** $\frac{5}{8} - \frac{1}{8} = \frac{1}{2}$ **17.** $\frac{5}{2} × \frac{3}{2} = 3\frac{3}{4}$ **18.** $\frac{3}{2} ÷ \frac{5}{2} = \frac{3}{5}$

19. What is the tenth number of this sequence? 2, 4, 6, 8, . . .
20

20. 40,200 ÷ 25 =
1608

21. 348 × 67 =
23,316

22. 48 − 84 =
⁻36

23. A meter is about one **big** step. About how many meters high is a door?   2 m

24. Five of the 30 students in the class were absent. What fraction of the class was absent? (Reduce.) $\frac{1}{6}$

25. To what number on the line is the arrow pointing? 4

LESSON
34

Decimal Fractions

One way of naming parts of whole amounts is with common fractions. Common fractions have both the numerator and denominator written out, like $\frac{1}{2}$ or $\frac{3}{10}$. Another way of naming parts of wholes is with **decimal fractions**. Decimals are actually fractions with denominators of 10, 100, 1000, etc. The denominator of a decimal fraction is not written out but is indicated by the place value of the decimal number.

One place after the decimal point is the tenths' place.

$$0.\underline{1} \text{ means } \frac{1}{10}$$

Two places after the decimal point is the hundredths' place.

$$0.1\underline{2} \text{ means } \frac{12}{100}$$

Three places after the decimal point is the thousandths' place.

$$0.12\underline{3} \text{ means } \frac{123}{1000}$$

example 34.1   Write 0.23 as a common fraction.

solution   Two places after the decimal point indicates hundredths, so 0.23 can be written $\frac{23}{100}$.

example 34.2   Write $\dfrac{3}{10}$ as a decimal.

solution   The denominator 10 can be indicated by one place after the decimal point (.__). In the tenths' place we write the 3. This gives us .3, which we write as **0.3**.

practice   Write as a common fraction.   Write as a decimal fraction.

a. 0.1 $\frac{1}{10}$ **g.** $\dfrac{7}{10}$ 0.7   **j.** $\dfrac{1}{100}$ 0.01

b. 0.3 $\frac{3}{10}$

c. 0.21 $\frac{21}{100}$ **h.** $\dfrac{13}{100}$ 0.13   **k.** $\dfrac{4}{100}$ 0.04

d. 0.321 $\frac{321}{1000}$

e. 0.03 $\frac{3}{100}$ **i.** $\dfrac{125}{1000}$ 0.125   **l.** $\dfrac{5}{1000}$ 0.005

f. 0.009 $\frac{9}{1000}$

problem set 34

1. When the sixth multiple of 5 is subtracted from the eighth multiple of 5, what is the difference?   10

2. Mom wants to triple a recipe for cheesecake. If the recipe calls for 8 ounces of cream cheese, how much must she put in?   24 oz.

3. What time is two and one-half hours after 10:40 A.M.?   1:10 P.M.

*4. Write 0.23 as a common fraction. $\frac{23}{100}$

*5. Write $\dfrac{3}{10}$ as a decimal.   0.3

6. Write $(6 \times 100) + (5 \times 1) + (3 \times \frac{1}{10})$ in standard notation.   605.3

7. Which digit is in the ones' place in 42,876.39?   6

8. How many **square millimeters** is the area of the square?   144 sq. mm

12 mm

9. How many **millimeters** is the perimeter of the square?   48 mm

10. What is the least common multiple of 6 and 8?   24

11. $0.55 + $0.56 = $1.11   **12.** Compare: $\dfrac{1}{10} \bigcirc 0.1$ =

13. Estimate the quotient when 898 is divided by 29.   30

14. Round 36,847 to the nearest hundred.   36,800

15. $\square \times 6 = 144$ **16.** $\dfrac{3}{8}$ of 24 = 9   **17.** $6 - 60 =$
   24 -54

Write each answer in simplest form.

18. $\dfrac{3}{8} + \dfrac{3}{8} = \frac{3}{4}$ **19.** $\dfrac{11}{12} - \dfrac{1}{12} = \frac{5}{6}$ **20.** $\dfrac{5}{4} \times \dfrac{3}{2} = 1\frac{7}{8}$ **21.** $\dfrac{1}{2} \div \dfrac{1}{3} = 1\frac{1}{2}$

22. Which number is missing? 6, 12, ___, 24, 30   18

23. $437 \times 86 = 37,582$ **24.** $5225 \div 12 = 435 \text{ r}5$

25. To which number is the arrow pointing? -6

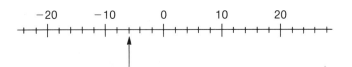

**LESSON
35**

Reading and Writing Decimal Numbers

reading decimals We read a decimal number the same way we read a whole number, **and then we say the place value of the last digit.** We read 0.23 by saying "twenty-three" and then we say "hundredths" because the last digit is in the hundredths' place.

example 35.1 Read: 0.023

solution **"Twenty-three thousandths"**

reading mixed decimals

To read mixed decimal numbers like 20.04 we read the whole number part, then we say "and" for the decimal point, and then we read the decimal fraction.

example 35.2 Read: 20.04

solution **"Twenty and four hundredths."**

writing decimals

When writing decimal numerals, we look for the place value of the last digit. That tells us how many decimal places are in the numeral. Then we fit the number back into that place.

example 35.3 Write the numeral for "twelve thousandths."

solution First look at "thousandths." Thousandths is three places after the decimal point (.___). We fit the twelve back into the last two places (._12) and then fill the empty space with 0 and we have .012, which we write as **0.012**.

practice (See Practice Set N in the Appendix.)

Use words to write these numbers. †

a. 0.123
b. 0.05
c. 0.015
d. 0.001
e. 0.26
f. 0.01

g. 1.2
h. 10.2
i. 12.04
j. 1.234
k. 10.1
l. 100.01

Use digits to write these numbers.

m. Five hundredths 0.05
n. Eleven thousandths 0.011
o. One and two tenths 1.2
p. Ten and five hundredths 10.05
q. Twenty three thousandths 0.023
r. Twenty and five tenths 20.5

problem set 35

1. What is the product of three-fourths and three-fifths? $\frac{9}{20}$

2. Bugs planted 360 carrot seeds in his garden. Three fourths of them grew. How many carrots grew? 270

3. Jan's birthday cake must bake for 2 hours and 15 minutes. If it is put into the oven at 11:45 A.M., at what time will it be done? 2 P.M.

*4. Write 0.023 in word form. twenty-three thousandths

*5. Write 20.04 in word form. twenty and four hundredths

† **a.** one hundred twenty-three thousandths **b.** five hundredths **c.** fifteen thousandths **d.** one thousandth **e.** twenty-six hundredths **f.** one hundredth **g.** one and two tenths **h.** ten and two tenths **i.** twelve and four hundredths **j.** one and two hundred thirty-four thousandths **k.** ten and one tenth **l.** one hundred and one hundredth

*6. Write the numeral for "twelve thousandths."   0.012

7. Write 0.17 as a fraction. $\frac{17}{100}$

8. Write $(5 \times 1000) + (6 \times 100) + (4 \times 10)$ in standard notation.   5640

9. Which digit in 1.23 is in the same place as the 5 in 0.456?   3

10. What is the area of the rectangle?   200 sq. mm

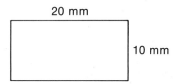

20 mm

10 mm

11. What is the perimeter of the rectangle?   60 mm

12. There are 100 centimeters in a meter. How many centimeters are in 10 meters?   1000 cm

13. Which number is between 2000 and 3000?   (b) 2639
(a) 3121   (b) 2639   (c) 1989   (d) 3163

14. A meter is about one **big** step. About how many meters wide is a door?   1

15. $\frac{3}{5} + \frac{2}{5} =$ 1   **16.** $\frac{5}{8} - \frac{5}{8} =$ 0   **17.** $\frac{2}{3} \times \frac{3}{4} = \frac{1}{2}$ **18.** $\frac{2}{5} \div \frac{2}{5} =$ 1

19. Convert $\frac{20}{6}$ to a mixed number, then reduce the fraction.
$3\frac{1}{3}$

20. $\frac{5}{6}$ of 24 = 20

21. $\frac{\square}{4} = 5$ 20

22. Compare: $\frac{3}{3} \bigcirc 1 + \frac{2}{2}$ <

23. What is the sum of the first six positive odd numbers? 36

24. $(16 \times 18) \div 12 = 24$

25. To which number on the line is the arrow pointing? -14

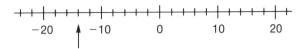

LESSON 36

Decimals Chart

Two ways of writing parts of whole numbers are with common fractions (like $\frac{1}{2}$) and with decimal fractions (like 0.5). Arithmetic is easy to do with decimal fractions because decimal number arithmetic is just like whole number arithmetic except for the decimal point. In decimal number arithmetic, you must keep track of the decimal point. The chart below will help you keep track of the decimal point. The six rules in the chart will help you to learn decimal arithmetic quickly. You will be asked to draw this chart many times to help remember the rules. The rules will be explained in detail in the lessons that follow.

The Decimals Chart

+ −	×	÷ BY WHOLE	÷ BY DECIMAL
Line up.	× then count.	Up.	Over, over, up.
1. Pin a decimal point on the back of a whole number. 2. Fill empty places with zero.			

Across the top of the chart are the four operation signs ($+$, $-$, $\times$, and $\div$).

Below each sign is the rule to follow when working that kind of problem. (There are two kinds of division problems so there are two different rules.)

The bottom of the chart contains two general rules which may be needed to solve different kinds of problems.

problem set 36   Draw the decimals chart.

1. The average of two numbers is 10. If one of the numbers is 7, what is the other number?   13

2. Jack accidentally sat on his lunch and smashed $\frac{3}{4}$ of his sandwich. What fraction of his sandwich was not smashed? $\frac{1}{4}$

3. A mile is 5,280 feet. There are 3 feet in a yard. How many yards are in a mile?   1760 yd.

4. Which digit in 23.47 has the same place value as the 6 in 516.9?   3

5. Write 1.3 in word form.   one and three tenths

6. Write the decimal numeral five hundredths.   0.05

7. Write 0.31 as a fraction. $\frac{31}{100}$

8. Write $(4 \times 100) + (3 \times 1) + (2 \times \frac{1}{100})$ in standard notation.   403.02

9. Which digit in 4.375 is in the tenths' place?   3

10. How many 1-inch square tiles are needed to cover this square?   1296 sq.-in. tiles

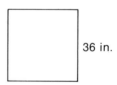

36 in.

11. What is the perimeter of the square?   144 in.

12. $3\frac{1}{4} + 2\frac{1}{4} =$ $5\frac{1}{2}$ 13. $3 - 1\frac{1}{4} =$ $1\frac{3}{4}$ 14. $3\frac{1}{3} + 2\frac{2}{3} =$ 6

15. $\frac{3}{4}$ of 28 =   21   16. $\frac{3}{4} \times \frac{4}{6} =$ $\frac{1}{2}$ 17. $\frac{5}{6} \div \frac{1}{2} =$ $1\frac{2}{3}$

18. What is the average of 42, 57, and 63? 54

19. The factors of 6 are 1, 2, 3, and 6. List the factors of 20.
1, 2, 4, 5, 10, 20

20. What is the least common multiple of 9 and 6? 18

21. $(10{,}000 - 1234) \div 18 = $ 487 **22.** $375 - 512 = $ −137

23. Round 58,742,177 to the nearest million. 59,000,000

24. Estimate the product of 823 and 680. 560,000

25. How many **millimeters** long is the line segment?
(1 cm = 10 mm) 50 mm

```
cm   1    2    3    4    5    6    7    8    9
   └─┴────┴────┴────┴────┴────┴────┴────┴────┘
```

LESSON
37

Adding and Subtracting Decimals: "Line Up"

"You can't add apples and oranges."

This saying means you cannot add **unlike** things. You can only add **like** things. When we add numbers we can only add digits with the same place value—ones with ones, tens with tens, and so on. When we add whole numbers we line up the ending digits because they are always in the ones' place, and lining up the ones' place will line up all the other places. However, lining up the ending digits of decimal numbers may not properly line up all the digits. We must use another method for decimal numbers. **We line up decimal numbers for addition or subtraction by lining up the decimal points.** Remember the rule: Line up the decimal points. **The decimal point in the answer goes under the other decimal points.**

example 37.1 Add: $3.4 + 0.26 + 0.3 = $

solution   Line up the decimal points in the problem
and add. The decimal point in the answer
is placed just below the other decimal
points.

$$\begin{array}{r} 3.4 \\ 0.26 \\ + \, 0.3 \\ \hline \mathbf{3.96} \end{array}$$

example 37.2   Subtract: $4.56 - 2.3 =$

solution   Line up the decimal points in the problem
and subtract.

$$\begin{array}{r} 4.56 \\ - \, 2.3 \\ \hline \mathbf{2.26} \end{array}$$

practice   Add or subtract. Remember to line up the decimal points.

a. $3.46 + 0.2 =$ 3.66   **f.** $8.28 - 6.1 =$ 2.18

b. $0.735 + 0.21 =$ 0.945   **g.** $0.543 - 0.21 =$ 0.333

c. $0.43 + 0.1 + 0.413 =$ 0.943   **h.** $0.30 - 0.27 =$ 0.03

d. $0.6 + 0.7 =$ 1.3   **i.** $1.00 - 0.24 =$ 0.76

e. $0.9 + 0.12 =$ 1.02   **j.** $1.23 - 0.4 =$ 0.83

problem set 37

1. What is the reciprocal of three eighths? $\frac{8}{3}$

2. Penny broke 8 pencils on her math test. She broke half as many on her spelling test. How many did she break in all?
12

3. What number must be added to three hundred seventy-five to total one thousand?   625

*4. $3.4 + 0.26 + 0.3 =$ 3.96   *5. $4.56 - 2.3 =$ 2.26

6. $\$0.37 + \$0.23 + \$0.48 =$ $1.08   7. $\$5 - 5¢ =$ $4.95

8. What is the next number? 1, 10, 100, 1000, ___   10,000

9. Each side of a square is 100 cm long. How many tiles 1 cm on each edge are needed to cover the area?   10,000 tiles

10. Which digit is in the ten-millions' place in 1,234,567,890?
3

11. Three of these numbers are equal. Which number is different?   (d) 0.01

(a) $\dfrac{1}{10}$ (b) 0.1   (c) $\dfrac{10}{100}$ (d) 0.01

12. Estimate the product of 29, 42, and 39. 48,000

13. $10,203 \div 10 =$ **14.** $32,100 \div 30 =$ **15.** $10,000 - 345 =$
     1020 r3   1070   9655

16. $\dfrac{3}{4} + \dfrac{3}{4} = 1\dfrac{1}{2}$ **17.** $3 - 1\dfrac{3}{5} = 1\dfrac{2}{5}$ **18.** $\dfrac{2}{2} \div \dfrac{3}{3} = 1$

19. $1\dfrac{1}{3} + 2\dfrac{1}{3} + 3\dfrac{1}{3} = 7$

20. Compare: $100 \div 6 \bigcirc 100 \div 7$ >

21. Convert the improper fraction $\dfrac{100}{7}$ to a mixed number.
     $14\frac{2}{7}$

22. What is the average of 90 lb., 84 lb., and 102 lb.?   92 lb.

23. What is the least common multiple of 4 and 5?   20

24. The temperature changed from $11°$ at noon to $-4°$ at 8:00 P.M.. How many degrees did the temperature drop?
     $15°$

25. To what point on the scale is the arrow pointing? $6\frac{3}{4}$

26. Draw the decimals chart.

LESSON
38

"Pin Decimal Point on Back of Whole Number"

Almost everyone knows where to pin the tail on the donkey. To be successful in a game of "pin the tail," you must find the end of the donkey. To be successful at decimal arithmetic, you must be able to find the end of the whole number. This is where the decimal point belongs. The decimal point shows where the whole number part of a number ends. The decimal point also shows where the fractional part of a number begins.

When we add and subtract decimal numbers, we line up the decimal points. This lets us add and subtract digits that have the same place values. **To help us line up whole numbers, we first "pin" the decimal point to the end of the whole number.** This decimal point is lined up with the decimal points in other numbers before we add or subtract.

example 38.1   Add: $3 + 1.2 =$

solution   To add like place values we must line up the decimal points. Since the 3 does not have a decimal point, we can "pin" one on the end. Then we line up the decimal points and add.

$$\begin{array}{r} 3. \\ + 1.2 \\ \hline \mathbf{4.2} \end{array}$$

practice
(See Practice Set O in the Appendix.)

a. $4 + 2.1 = 6.1$
b. $3 + 0.4 = 3.4$
c. $12.5 + 10 = 22.5$
d. $0.23 + 4 + 3.7 = 7.93$

e. $4.3 - 2 = 2.3$
f. $43.2 - 5 = 38.2$
g. $6.3 - 6 = 0.3$
h. $145.75 - 25 = 120.75$

problem set 38   Draw the decimals chart.

1. What is the largest number which can divide evenly into both 54 and 45?   9

2. Roberto began saving $3 each week for summer camp, which costs $126. How many weeks will it take to save that amount?   42 weeks

3. Ghandi was born in 1869. How old was he when he was assassinated in 1948?   79 years

*4. $3 + 1.2 = 4.2$ 5. $3.6 + 4 = 7.6$ 6. $5.63 - 1.2 = 4.43$

7. $5.376 + 0.24 = 5.616$ 8. $4.75 - 0.6 = 4.15$ 9. $\$4 - 4\cent = \$3.96$

10. Write 0.47 as a fraction. $\frac{47}{100}$

11. Write $(9 \times 1000) + (4 \times 10) + (3 \times 1) + (4 \times \frac{1}{10})$ in standard notation.   9043.4

12. Which digit is in the hundredths' place in $123.45?   5

13. The perimeter of a square is 100 inches. How long is each side?   25 in.

14. What is the least common multiple of 2, 3, and 4?   12

15. $1\frac{2}{3} + 2\frac{2}{3} =$ **16.** $5 - 1\frac{1}{4} =$ **17.** $\frac{3}{4} \times \frac{4}{5} =$ **18.** $\frac{1}{4} \div \frac{3}{4} =$
$4\frac{1}{3}$ $3\frac{3}{4}$ $\frac{3}{5}$ $\frac{1}{3}$

19. Instead of dividing $\frac{5}{6}$ by $\frac{2}{3}$, you could multiply $\frac{5}{6}$ by what?
$\frac{3}{2}$

20. Six of the nine players got on base. What fraction of the players got on base? (Always reduce.) $\frac{2}{3}$

21. List the factors of 30.   1, 2, 3, 5, 6, 10, 15, 30

22. What number is halfway between 5,987 and 3,143?   4565

23. Round 186,497 to the nearest thousand.   186,000

24. $123 - 486 = -363$ **25.** $\dfrac{22 + 23 + 24}{23} = 3$

LESSON 39

Multiplying Decimals— "Multiply, Then Count"

The rule for multiplying decimal numbers is to "multiply, then count." We multiply decimal numbers the same way we multiply whole numbers. We do not attempt to line up the decimal points. That rule is only for adding and subtracting. We **ignore** decimal points until we have an answer. Then we count the total number of digits to the right of the decimal points in the problem. We place the decimal point in the answer so that there are the same total of digits to the right of the decimal point in the answer.

$$
\begin{array}{r}
1 . 2 \ (1) \\
\times \quad 3 \\
\hline
3 . \underline{6} \ (1)
\end{array}
\qquad
\begin{array}{r}
0 . 1 \ 2 \ (2) \\
\times \quad 3 \\
\hline
0 . \underline{3} \ \underline{6} \ (2)
\end{array}
\qquad
\begin{array}{r}
1 \ 2 \\
\times \ 0 . 3 \ (1) \\
\hline
3 . \underline{6} \ (1)
\end{array}
\qquad
\begin{array}{r}
1 . 2 \} \\
\times \ \ 0 . 3 \} \ (2) \\
\hline
0 . \underline{3} \ \underline{6} \ \ (2)
\end{array}
$$

example 39.1   Multiply: 0.25×0.7

solution   We set up like a whole number problem and multiply. Then we count the number of digits to the right of the decimal points in the problem. In this problem, we count 2 plus 1 equals 3. Then we put the decimal point 3 places from the right end. We write .175 as **0.175**.

$$
\begin{array}{r}
0 . 2 \ 5 \} \\
\times \quad 0 . 7 \} \ (3) \\
\hline
. \underline{1} \ 7 \ \underline{5}
\end{array}
$$

practice   "Multiply, then count."

a. $15 \times 0.3 =$ 4.5   **e.** $2.5 \times 0.5 =$ 1.25
b. $1.5 \times 3 =$ 4.5   **f.** $0.25 \times 0.5 =$ 0.125
c. $0.15 \times 3 =$ 0.45   **g.** $1.5 \times 1.5 =$ 2.25
d. $1.5 \times 0.3 =$ 0.45   **h.** $0.25 \times 2.5 =$ 0.625

problem set 39   Draw the decimals chart.

1. Mount Everest, the world's tallest mountain, is twenty-nine thousand, twenty-eight feet high. Write that number. 29,028

2. There are three feet in a yard. How many yards high is Mt. Everest?   9676 yd.

3. Bam Bam says his pet dinosaur weighs $\frac{3}{4}$ as much as a garbage truck. If the truck weighs 12 tons, how much does his dinosaur weigh?   9 tons

*4. $0.25 \times 0.7 =$
0.175

5. $1.8 \times 0.9 =$
1.62

6. $63 \times 0.7 =$
44.1

7. $1.23 + 4 + 0.5 =$
5.73

8. $12.34 - 5.6 =$
6.74

9. $\$3 - 3¢ =$
$2.97

10. Write the decimal number ten and two-tenths.   10.2

11. Write the decimal number twenty-three thousandths.
0.023

12. Write the standard numeral for $(6 \times 100) + (4 \times 10) + (3 \times \frac{1}{10})$.   640.3

13. Which digit in 3.675 has the same place value as the 4 in 14.28?   3

14. The perimeter of a square is 100 inches. How many square tiles 1 inch on each edge are needed to cover its area?
625 sq. in.

15. What is the least common multiple (LCM) of 2, 3, and 6?
6

16. Convert this improper fraction to a mixed number and reduce the fraction: $\frac{20}{8}$. $2\frac{1}{2}$

17. $\left(\dfrac{1}{3} + \dfrac{2}{3}\right) - 1 =$ 0   **18.** $\dfrac{3}{5} \times \dfrac{2}{3} =$ $\dfrac{2}{5}$ **19.** $\dfrac{2}{5} \div \dfrac{1}{2} =$ $\dfrac{4}{5}$

20. A pie was cut into six equal slices. Two slices were eaten. What fraction of the pie is left? (Always reduce when possible.) $\frac{2}{3}$

21. List the factors of 33.   1, 3, 11, 33

22. On Tim's last four papers he had 21, 24, 23, and 20 right. He averaged how many right on these papers?   22

23. Estimate the quotient when 7987 is divided by 39.   200

24. Compare: $365 - 364 \bigcirc 364 - 365$ >

25. How long is the line segment?
$2\frac{3}{4}$ in.

LESSON
40

"Fill Empty Places with Zeros"

When subtracting, multiplying, and dividing decimal numbers we often find a decimal place with no digit in it, like these.

$$
\begin{array}{ccc}
& & 0.\_4 \\
0.5\_ & 0.2 & 3\,)\overline{0.12} \\
-0.32 & \times 0.3 & \\
\hline
& \overline{0.\_6} &
\end{array}
$$

A blank place means the number contains none of that place value. We may and often **must** fill an empty place with zero.

subtraction In order to subtract it is sometimes necessary to add zeros to the top number.

$$
\begin{array}{r}
0.5\,0 \\
-0.3\,2 \\
\hline
\end{array}
$$

example 40.1 $3 - 0.4 =$

solution Pin the decimal on the back of the whole
number. Line up the decimal points. Fill
the empty place with zero and subtract.

$$
\begin{array}{r}
3.0 \\
-0.4 \\
\hline
\mathbf{2.6}
\end{array}
$$

multiplication When multiplying, we must sometimes
insert zero(s) between the multiplication
answer and the decimal point.

$$
\begin{array}{r}
0.2 \\
\times 0.3 \\
\hline
0.06
\end{array}
$$

example 40.2 $0.12 \times 0.3 =$

solution Multiply, then count three places. Fill
the empty place with zero.

$$
\begin{array}{r}
0.1\,2 \\
\times 0.3 \\
\hline
\mathbf{0.0\,3\,6}
\end{array}
$$

practice
(See Practice
Set P in the
Appendix.)

a. $4.6 - 0.46 = 4.14$
b. $0.4 - 0.32 = 0.08$
c. $1 - 0.4 = 0.6$
d. $5 - 1.25 = 3.75$
e. $0.3 \times 0.3 = 0.09$

f. $0.1 \times 0.2 \times 0.3 = 0.006$
g. $0.12 \times 0.12 = 0.0144$
h. $0.01 \times 1 = 0.01$
i. $0.1 \times 0.1 = 0.01$

problem set 40 Draw the decimals chart.

1. What is the sum of one hundred twenty-four thousand and ninety thousand, six hundred five?   214,605

2. The first slaves were taken to the colony of Virginia in 1619. African slave trade ended in 1871. How many years did the slave trade last?   252 years

3. White Rabbit is three and a half hours late for a very important date. If the time is 2:00 P.M., what was the time of his date?   10:30 A.M.

*4. $3 - 0.4 =$ 2.6   5. $1.2 - 0.12 =$ 1.08   6. $1 - 0.1 =$ 0.9

*7. $0.12 \times 0.3 =$ 0.036   8. $0.01 \times 0.1 =$ 0.001   9. $4.8 \times 0.23 =$ 1.104

10. Use digits to write the number one and two hundredths. 1.02

11. Write the standard numeral for $(6 \times 10,000) + (8 \times 100)$. 60,800

12. A square is covered by 64 tiles. How many tiles are along each side?   8 tiles

13. What is the least common multiple (LCM) of 2, 4, and 8? 8

14. $6\dfrac{2}{3} + 4\dfrac{2}{3} = 11\dfrac{1}{3}$ 15. $5 - 3\dfrac{3}{8} = 1\dfrac{5}{8}$ 16. $\dfrac{5}{8} \times \dfrac{2}{3} = \dfrac{5}{12}$

17. $\dfrac{3}{5} \div \dfrac{1}{2} = 1\dfrac{1}{5}$ 18. Compare: $\dfrac{1}{2} \times \dfrac{2}{2} \bigcirc \dfrac{1}{2} \times \dfrac{3}{3}$ =

19. $1000 - \square = 567$ 433

20. Eighteen of the thirty students in the class received A's. What fraction of the class received A's? $\frac{3}{5}$

21. How many whole numbers are factors of 100?   9

22. $\dfrac{100 + 200 + 300 + 400}{25} = 40$

23. Round $4167 to the nearest hundred dollars. $4200

The circle gives us some information about the test scores of some students who took a test. This type of graph is sometimes called a **circle graph**.

Use this graph of grades received on a math test to answer questions 24 and 25.

24. How many students took the test?
30

25. What fraction of the students received a grade of *C* on the test? $\frac{1}{6}$

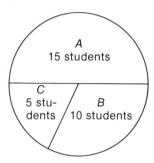

LESSON
41

The Name-Changer Machine

Have you ever wished you could change your name? Numbers can. Most people have only two or three names, but a number never runs out of names. All a number needs to do is jump into the name-changer machine to get another one of its millions of names. Here's how the name-changer machine works.

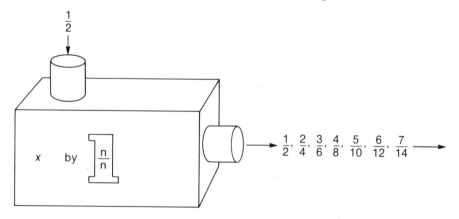

The name changer multiplies the number by 1 and changes its form but does not change its value. The name changer multiplies the number by one of these special forms of the number 1.

$$\frac{1}{1}, \frac{2}{2}, \frac{3}{3}, \frac{4}{4}, \frac{5}{5}, \frac{6}{6}, \frac{7}{7}, \text{etc.}$$

If you multiply a number by one of these special ways to write 1, you will get a different name for the number you multiplied. If we begin with the number $\frac{1}{2}$, we can change its name. We show three new names for $\frac{1}{2}$ here.

$$\frac{1}{2} \times \boxed{\frac{2}{2}} = \frac{2}{4} \qquad \frac{1}{2} \times \boxed{\frac{3}{3}} = \frac{3}{6} \qquad \frac{1}{2} \times \boxed{\frac{5}{5}} = \frac{5}{10}$$

example 41.1 By what name for 1 must $\frac{1}{2}$ be multiplied to result in a fraction with a 6 in the denominator?

solution To get a 6 in the denominator we must multiply by the name of 1 called $\frac{3}{3}$ because

$$\frac{1}{2} \times \boxed{} = \frac{?}{6}$$

$$\frac{1}{2} \times \frac{3}{3} = \frac{3}{6}$$

example 41.2 $\dfrac{1}{2} = \dfrac{?}{8}$

solution To change halves into eighths, multiply by the name of 1, $\frac{4}{4}$.

$$\frac{1}{2} \times \frac{4}{4} = \frac{4}{8}$$

practice **a.** $\dfrac{1}{2} \times \, ? = \dfrac{6}{12}$ $\dfrac{6}{6}$

d. $\dfrac{3}{4} \times \, ? = \dfrac{6}{8}$ $\dfrac{2}{2}$

g. $\dfrac{1}{4} = \dfrac{?}{12}$ 3

b. $\dfrac{1}{3} \times \, ? = \dfrac{4}{12}$ $\dfrac{4}{4}$

e. $\dfrac{1}{2} = \dfrac{?}{10}$ 5

h. $\dfrac{2}{3} = \dfrac{?}{9}$ 6

c. $\dfrac{2}{3} \times \, ? = \dfrac{4}{6}$ $\dfrac{2}{2}$

f. $\dfrac{1}{3} = \dfrac{?}{6}$ 2

problem set 41 Draw the decimals chart.

1. The average of two numbers is 20. If one of the numbers is 24, what is the other number? 16

2. Our own galaxy, the Milky Way, may contain two hundred billion stars. Write that number. 200,000,000,000

3. The rectangular school yard is 120 yards long and 40 yards wide. How many square yards is its area?   4800 sq. yd.

*4. By what name for 1 must $\frac{1}{2}$ be multiplied to result in a fraction with a 6 in the denominator? $\frac{3}{3}$

*5. $\dfrac{1}{2} = \dfrac{?}{8}$ 4   **6.** $\dfrac{1}{2} = \dfrac{?}{10}$ 5

7. $4.32 + 0.6 + 11 =$ **8.** $6.3 - 0.54 =$ **9.** $0.15 \times 0.15 =$
    15.92   5.76   0.0225

10. What is the reciprocal of $\dfrac{6}{7}$? $\dfrac{7}{6}$

11. Which digit in 12,345 has the same place value as the 6 in 67.89? 4

12. What is the least common multiple of 3, 4, and 6? 12

13. $5\dfrac{3}{5} + 4\dfrac{4}{5} = 10\dfrac{2}{5}$ **14.** $6 - 4\dfrac{2}{3} = 1\dfrac{1}{3}$ **15.** $\dfrac{8}{3} \times \dfrac{1}{2} = 1\dfrac{1}{3}$

16. $\dfrac{8}{3} \div \dfrac{1}{2} = 5\dfrac{1}{3}$ **17.** $1 - \dfrac{1}{4} = \dfrac{3}{4}$ **18.** $\dfrac{10}{10} - \dfrac{5}{5} = 0$

19. Four dimes is what fraction of a dollar? $\frac{2}{5}$ **20.** List the factors of 35.
                                         1, 5, 7, 35

21. Alma's three scores were 12,643 and 9,870 and 14,261. What was her average score per game? 12,258

22. Estimate the quotient of $\dfrac{9176}{41}$. 225

23. How many doughnuts are in $\dfrac{2}{3}$ of a dozen? 8

24. What number is 100 less than 37? -63

25. How long is the line segment? $2\frac{1}{2}$ in.

LESSON
42

Measuring to the Nearest $\frac{1}{8}$ of an Inch

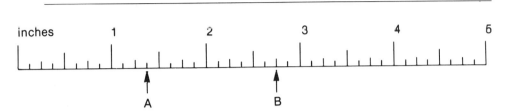

Looking at the scale above we see that the distance between the inches has been divided into 8 **spaces**. Each space is $\frac{1}{8}$ of an inch long. Arrow A is pointing to a mark which is 3 spaces beyond the 1-inch mark. The mark the arrow is pointing to then is the $1\frac{3}{8}$-inch mark. Arrow B is pointing to a mark 6 spaces beyond the 2-inch mark. It is pointing to the $2\frac{6}{8}$-inch mark. Remember, though, fractions which **can be** reduced **should be** reduced. The fraction $\frac{6}{8}$ reduces to $\frac{3}{4}$, so arrow B points to the $2\frac{3}{4}$-inch mark.

example 42.1 How long is this line segment?

solution The distance between the inches has been divided into 8 spaces. The line ends 5 spaces past the 1. The line is **$1\frac{5}{8}$ inches** long.

example 42.2 To which mark is the arrow pointing?

solution The arrow is pointing to the $1\frac{4}{8}$-inch mark, which reduces to **$1\frac{1}{2}$ inches**.

practice To which mark on the inch scale is each arrow pointing?

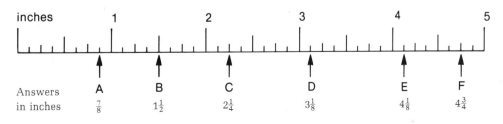

Answers in inches
A $\frac{7}{8}$ B $1\frac{1}{2}$ C $2\frac{1}{4}$ D $3\frac{1}{8}$ E $4\frac{1}{8}$ F $4\frac{3}{4}$

problem set 42   Draw the decimals chart.

1. What number must be added to six thousand, eighty-four to get a sum of ten thousand?   3916

2. One hundred fifty knights could sit at the Round Table. King Arthur saw that only one hundred twenty-eight of his knights were seated at the table. How many empty places were at the table?   22

3. Frank started running the marathon at 11:50 A.M. and finished 2 hours and 11 minutes later. At what time did he finish?   2:01 P.M.

4. $1 - \dfrac{3}{10} = \dfrac{7}{10}$

5. $6 - 1\dfrac{4}{5} = 4\dfrac{1}{5}$

6. $6\dfrac{3}{4} + 4\dfrac{1}{4} = 11$

7. $\dfrac{5}{8} \times \dfrac{1}{5} = \dfrac{1}{8}$

8. $\dfrac{3}{4} \div \dfrac{2}{3} = 1\dfrac{1}{8}$

9. $\dfrac{2}{3} \times \square = 1$ $\dfrac{3}{2}$

10. $\dfrac{2}{3} = \dfrac{?}{6}$ 4

11. $\dfrac{1}{2} = \dfrac{?}{6}$ 3

12. Compare: $\dfrac{2}{2} \bigcirc \dfrac{2}{2} \times \dfrac{2}{2}$ =

13. The temperature was 8° at midnight but dropped 15° by morning. What was the morning temperature? $-7°$

14. Write the decimal numeral for the number nine and twelve hundredths.   9.12

15. Round 67,492,384 to the nearest million.   67,000,000

16. $46.37 + 5.93 + 14 =$ 66.30

17. $12 - 1.43 =$ 10.57

18. $0.37 \times 0.26 =$ 0.0962

19. $0.4 \times 0.6 \times 0.8 =$ 0.192

20. A square room is covered with 64 square floor tiles 1 foot on each edge. What is the perimeter of the room?   32 ft.

21. $32 \overline{)1000}$ 31 r8

22. $53 - 530 =$ -477

23. What is the least common multiple (LCM) of 4, 6, and 8?
24

*24. How long is this line segment?
1$\frac{5}{8}$ in.

*25. To which mark is the arrow pointing?
1$\frac{1}{2}$ in.

inches   1   2

LESSON
43

Subtracting Fractions with "Borrowing"

Sometimes we need to regroup when we subtract whole numbers. In this example, we take 10 from the tens' column to get 13 in the ones' column. Then we subtract.

$$\begin{array}{r} 73 \\ -27 \\ \hline \end{array} \longrightarrow \begin{array}{r} \overset{6}{\cancel{7}}{}^{1}3 \\ -2\,7 \\ \hline 4\,6 \end{array} \quad \text{regrouped}$$

Sometimes we need to regroup when we subtract mixed numbers. In the next problem, we take 1 from 5 and write the 1 as $\frac{3}{3}$. Then we subtract.

$$\begin{array}{r} 5 \\ -1\frac{2}{3} \\ \hline \end{array} \longrightarrow \begin{array}{r} \overset{4}{\cancel{5}}+1 \\ -1\quad\frac{2}{3} \\ \hline \end{array} \longrightarrow \begin{array}{r} 4\frac{3}{3} \\ -1\frac{2}{3} \\ \hline 3\frac{1}{3} \end{array}$$

In the next problem, we take 1 from 5. In the fraction column we write 1$\frac{1}{3}$. Then we change 1$\frac{1}{3}$ to $\frac{4}{3}$ and subtract.

$$\begin{array}{r} 5\frac{1}{3} \\ -1\frac{2}{3} \\ \hline \end{array} \longrightarrow \begin{array}{r} \overset{4}{\cancel{5}}+1\frac{1}{3} \\ -1\frac{2}{3} \\ \hline \end{array} \longrightarrow \begin{array}{r} 4\frac{4}{3} \\ -1\frac{2}{3} \\ \hline 3\frac{2}{3} \end{array}$$

The 10 problems in the practice set below will give you practice in borrowing.

practice
(See Practice
Set U in the
Appendix.)

a. $4\frac{1}{3}$
$-1\frac{2}{3}$
$\overline{\ \ 2\frac{2}{3}}$

b. $3\frac{1}{5}$
$-1\frac{3}{5}$
$\overline{\ \ 1\frac{3}{5}}$

c. $4\frac{2}{5}$
$-2\frac{3}{5}$
$\overline{\ \ 1\frac{4}{5}}$

d. $6\frac{1}{4}$
$-5\frac{2}{4}$
$\overline{\ \ \frac{3}{4}}$

e. $5\frac{2}{4}$
$-1\frac{3}{4}$
$\overline{\ \ 3\frac{3}{4}}$

f. $7\frac{1}{6}$
$-1\frac{2}{6}$
$\overline{\ \ 5\frac{5}{6}}$

g. $9\frac{2}{6}$
$-3\frac{3}{6}$
$\overline{\ \ 5\frac{5}{6}}$

h. $4\frac{3}{8}$
$-1\frac{4}{8}$
$\overline{\ \ 2\frac{7}{8}}$

i. $5\frac{1}{8}$
$-2\frac{4}{8}$
$\overline{\ \ 2\frac{5}{8}}$

j. $7\frac{3}{12}$
$-4\frac{10}{12}$
$\overline{\ \ 2\frac{5}{12}}$

problem set 43

Draw the decimals chart.

1. What is the sum of the third multiple of 4 and the third multiple of 5?   27

2. Mt. Everest is 29,028 feet high. Mt. Whitney is 14,495 feet high. How much higher is Mt. Everest?   14,533 ft.

3. A mile is 5,280 feet. How many feet more than 5 miles high is Mt. Everest?   2628 ft.

4. $5\frac{1}{3}$
$-1\frac{2}{3}$
$\overline{\ \ 3\frac{2}{3}}$

5. $6\frac{2}{5}$
$-1\frac{3}{5}$
$\overline{\ \ 4\frac{4}{5}}$

6. $7\frac{1}{4}$
$-1\frac{3}{4}$
$\overline{\ \ 5\frac{1}{2}}$

7. What is the reciprocal of $\frac{5}{16}$? $\frac{16}{5}$

8. Write thirty-two thousandths as a decimal number.   0.032

9. $6\overline{)24,042}$ 4007

10. $10\overline{)3600}$ 360

11. $\dfrac{23+32}{55}=1$

12. Write the standard numeral for $(6 \times 10{,}000) + (4 \times 100) +$ (8×10).   60,480

13. $24\frac{2}{3} + 6\frac{2}{3} = 31\frac{1}{3}$ **14.** $\frac{7}{8} \div \frac{1}{2} = 1\frac{3}{4}$

15. What is the least common multiple of 2, 3, 4, and 6?   12

16. $6.74 + 0.285 + 4 = 11.025$ **17.** $0.4 - 0.33 = 0.07$

18. $1.6 \times 4.2 = 6.72$ **19.** $\frac{3}{4} = \frac{?}{8}$ 6   **20.** $1\frac{2}{3} = 1\frac{?}{12}$ 8

21. Find the average of 26, 37, 42, and 43.   37

22. Round 364,857 to the nearest thousand.   365,000

23. Write the next two numbers in this sequence. 1, 6, 4, 9, 7, 12, ___, ___.   10, 15

24. How many different whole numbers can divide 100 evenly?   9

25. How long is the line segment? $1\frac{7}{8}$ in.

LESSON
44

Dividing a Decimal Number by a Whole Number

When we divide a decimal number by a whole number, we put the decimal point in the answer (quotient) **straight up** from the decimal point in the decimal number.

$$3\overline{)4.2}^{\;\cdot} \qquad 3\overline{)0.24}^{\;\cdot}$$

Decimal division answers cannot be written with remainders. Instead we write zeros at the end of the decimal number and **keep dividing**.

example 44.1   3) 4.2

solution   The decimal point in the quotient is straight up from the decimal point in the dividend.

$$\begin{array}{r} 1.4 \\ 3\overline{)4.2} \\ \underline{3} \\ 1\,2 \\ \underline{1\,2} \\ 0 \end{array}$$

example 44.2   3) 0.24

solution   The decimal point in the quotient is straight up. We always put 0 in the empty places.

$$\begin{array}{r} 0.08 \\ 3\overline{)0.24} \\ \underline{24} \\ 0 \end{array}$$

example 44.3   5) 0.6

solution   The decimal point in the quotient is straight up. Note how we use 0's when needed to complete the division.

$$\begin{array}{r} 0.12 \\ 5\overline{)0.60} \\ \underline{5} \\ 10 \\ \underline{10} \\ 0 \end{array}$$

practice
(See Practice Set Q in the Appendix.)

a. 3) 4.5   1.5   e. 4) 0.2   0.05   i. 0.6 ÷ 4 = 0.15

b. 2) 0.14   0.07   f. 4) 0.3   0.075   j. 0.4 ÷ 5 = 0.08

c. 6) 0.012   0.002   g. 1.4 ÷ 7 = 0.2   k. 0.5 ÷ 4 = 0.125

d. 5) 0.7   0.14   h. 0.15 ÷ 5 = 0.03   l. 0.1 ÷ 4 = 0.025

problem set 44   Draw the decimals chart.

1. By what fraction must $\frac{5}{3}$ be multiplied to have a product equal to 1? $\frac{3}{5}$

2. How many twenty-dollar bills equal one thousand dollars?
50

3. Cindy made $\frac{2}{3}$ of her 24 shots at the basket. Each basket was worth 2 points. How many points did she make?
32 pt.

*4. 3) 4.2   1.4   *5. 3) 0.24   0.08   *6. 5) 0.6   0.12

7. What is the least common multiple (LCM) of 2, 4, 6, and 8?
24

8. $6 - 2\dfrac{3}{10} = 3\dfrac{7}{10}$ **9.** $5\dfrac{1}{5} - 2\dfrac{2}{5} = 2\dfrac{4}{5}$ **10.** $4\dfrac{3}{8} - 2\dfrac{5}{8} = 1\dfrac{3}{4}$

11. Estimate the product of 694 and 412.   280,000

12. $5.36 + 9 + 0.742 =$ 15.102   **13.** $3 - 1.56 =$ 1.44

14. $0.6 \times 0.7 \times 0.8 =$ 0.336   **15.** $0.46 \times 0.17 =$ 0.0782

16. Convert the improper fraction $\frac{40}{6}$ to a mixed number with the fraction reduced. $6\frac{2}{3}$

17. $9\dfrac{3}{5} + 2\dfrac{4}{5} = 12\dfrac{2}{5}$ **18.** $\dfrac{9}{4} \times \dfrac{1}{2} = 1\dfrac{1}{8}$ **19.** $\dfrac{3}{5} \div \dfrac{5}{3} = \dfrac{9}{25}$

20. Write a fraction equal to $\dfrac{5}{6}$ which has a denominator of 12.
$\dfrac{10}{12}$

21. The perimeter of a square is 24 feet. The area of the square is how many square feet?   36 sq. ft.

22. What number is halfway between 6375 and 4123?   5249

23. Write 0.27 as a fraction. $\dfrac{27}{100}$

24. How many years were there from 25 B.C. to 16 A.D.?
40 years

25. How long is the line segment?
$2\frac{3}{4}$ in.

LESSON
45

Dividing by a Decimal Number— Over, Over, Up

When the **divisor** of a division problem is a decimal number, we take an extra step to change the divisor to a whole number before we divide. We remember that the decimal point in a whole number is at the end of the number. First, we move the decimal point in the divisor to the end of the divisor. Then we move the decimal point in the dividend the same number of places. Next, we place the decimal point in the answer "straight up" from the decimal point in the dividend.

example 45.1 $0.3\overline{)0.36}$

solution   Since the divisor (0.3) is a decimal number we follow the rule "**over, over, up**" before dividing.

$$0.3\overline{)0.36}$$
over   over   up
1 . 2

example 45.2 $0.04\overline{)0.8}$

solution   To make the divisor a whole number we must move the decimal point over two places. Then we move the decimal point in the dividend over two places, adding zeros in the dividend as necessary.

$$0.04\overline{)0.80.}$$
over 2   over 2   up
2 0.

example 45.3 $0.3\overline{)6}$

solution   We pin a decimal on the back of the whole number. Then we follow the rule "over, over, up," adding zeros as necessary. Then we divide.

$$0.3\overline{)6.0.}$$
over 1   over 1   up
2 0.

practice
(See Practice
Set R in the
Appendix.)

a. $0.4\overline{)0.24}$ 0.6
b. $0.3\overline{)0.012}$ 0.04
c. $0.05\overline{)0.25}$ 5
d. $0.04\overline{)1.2}$ 30
e. $0.06\overline{)0.3}$ 5
f. $0.002\overline{)0.4}$ 200

g. $0.3\overline{)9}$ 30
h. $0.4\overline{)2}$ 5
i. $0.5\overline{)25}$ 50
j. $0.3 \div 0.03 = 10$
k. $2 \div 0.5 = 4$
l. $1.2 \div 0.01 = 120$

problem
set 45

Draw the decimals chart.

1. When a fraction with a numerator of 30 and a denominator of 8 is converted to a mixed number and reduced, what is the result? $3\frac{3}{4}$

2. Normal body temperature is 98.6° on the Fahrenheit scale. A person with a temperature of 100.2° would have a temperature how many degrees above normal?   1.6°

3. Four and twenty blackbirds is how many dozen?   2 doz.

*4. $0.3\overline{)0.36}$ 1.2   ***5.** $0.04\overline{)0.8}$ 20   ***6.** $0.3\overline{)6}$ 20

7. $5\overline{)6.35}$ 1.27   **8.** $4\overline{)0.5}$ 0.125   **9.** $8\overline{)1.0}$ 0.125

10. $9 - 3\dfrac{5}{8} = 5\dfrac{3}{8}$ **11.** $16\dfrac{1}{4} - 4\dfrac{3}{4} = 11\dfrac{1}{2}$ **12.** $2\dfrac{3}{8} - 1\dfrac{7}{8} = \dfrac{1}{2}$

13. $5.63 + 26.9 + 12 =$ 44.53   **14.** $1 - 0.235 =$ 0.765

15. $3.7 \times 0.25 =$ **16.** $\dfrac{3}{4} = \dfrac{?}{8}$ 6   **17.** $2\dfrac{3}{4} = 2\dfrac{?}{12}$
0.925   9

18. What is the least common multiple of 3, 4, and 8?   24

19. Compare: $1000 - 432 \bigcirc 1000 - 433$ >

20. Which digit is in the thousandths' place in 1,234.5678?
7

21. Write a fraction equal to 1 which has a denominator of 8.
$\dfrac{8}{8}$

22. Estimate the quotient when 3967 is divided by 48.   80

23. The area of a square is 100 square centimeters. How long is each side?   10 cm

24. There are 100 centimeters in 1 meter and 1000 meters in 1 kilometer. How many centimeters are in 2 kilometers?
200,000 cm

25. How long is the line segment?
$\frac{7}{8}$ in.

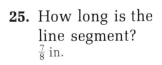

LESSON
46

Simplifying Decimal Numbers

Part of doing mathematics is writing answers in their simplest form. We simplify common fractions by reducing. We can also simplify decimal numbers.

The number 003.00 is not in its simplest form. As we look at the number we see a 3 in the ones' place. The simplest way to write the number which has a 3 in the ones' place is to just write 3 with no zeros or decimal point.

Generally, decimal numbers can be simplifed in three ways.

1. Remove zeros at the end of decimal numbers.
2. Remove the decimal point if no digits follow it.
3. Remove zeros in front of whole numbers.

Using our example: 003.00

1. The ending zeros may be removed, leaving

003.

2. With no digits after the decimal point, the point may be removed, leaving

003

3. Zeros before the whole number may be removed, leaving

3

example 46.1   Simplify: 01.20

solution   Remove zeros before whole number and after decimal number, leaving **1.2**

example 46.2   Simplify: 0.10

solution   It is becoming customary to write a digit in the ones' place even if that digit is zero. Removing only the zero at the end of the decimal part of the number leaves **0.1**

practice   Simplify these decimal numbers.

a. 02.0 2	**d.** 200.0 200	**g.** 1.020 1.02
b. 0.20 0.2	**e.** 020.0 20	**h.** 20.00 20
c. 00.2 0.2	**f.** 01.02 1.02	**i.** 020.0 20

problem   Draw the decimals chart.
set 46

1. The first positive odd number is 1. The second is 3. What is the tenth positive odd number?   19

2. Giant tidal waves can travel 500 miles per hour. How long would it take a tidal wave traveling at that speed to cross 3000 miles of ocean?   6 hr.

3. José bought Carmen one dozen red roses, two for each month he had known her. How long had he known her?
6 mos.

*4. Simplify: 01.20   1.2   ***5.** Simplify: 0.10   0.1

6. $0.3 \overline{)0.54}$ 1.8   **7.** $0.05 \overline{)3.7}$ 74   **8.** $0.6 \overline{)9}$ 15

9. $12 \overline{)0.18}$ 0.015   **10.** $10 \overline{)12.30}$ 1.23   **11.** $3.6 \times 12 = 43.2$

12. What is $\frac{5}{8}$ of 32?   20

13. Three of the twelve months begin with the letter J. What fraction of the months begin with J? $\frac{1}{4}$

14. $10 - 5\frac{11}{12} = 4\frac{1}{12}$ **15.** $6\frac{1}{5} - 3\frac{3}{5} = 2\frac{3}{5}$ **16.** $8\frac{3}{4} + 5\frac{3}{4} = 14\frac{1}{2}$

17. $\frac{5}{3} \times \frac{5}{4} = 2\frac{1}{12}$ **18.** $\frac{5}{3} \div \frac{5}{4} = 1\frac{1}{3}$ **19.** $\frac{3}{5} = \frac{?}{20}$ 12

20. What is the average of 492, 376, and 416?   428

21. List the factors of 20.   1, 2, 4, 5, 10, 20

22. What year was 10 years before 6 A.D.?   5 B.C.

23. What number is equal to "threescore and ten"? (Remember, a score is 20.)   70

24. Use the chart to find out how many more days it takes Mars to go around the sun than it takes Earth to go around the sun.
322 days

25. In the time it takes Mars to travel around the sun once, Venus travels around the sun about how many times?   3

PLANET	EARTH DAYS TO ORBIT THE SUN
Mercury	88
Venus	225
Earth	365
Mars	687

LESSON 47

Equivalent Decimals

The purpose of a decimal point is to indicate place value. The place to the left of the decimal point is always the ones' place. The place to the right is always the tenths' place. The place value of the 3 in 0.3 is exactly the same value as the 3 in 0.30 or 0.300. All of these show a 3 in the tenths' place. The extra zeros following the 3 do not change the place value of the 3.

We are used to seeing extra zeros after the decimal point in our money system. $3 and $3.00 are both ways of writing "three dollars," and of course $3 **equals** $3.00. We must also get used to the idea that 0.3 equals 0.30, which equals 0.300 and so on. In the last lesson we saw that the simplest way to write 0.300 is 0.3. In the next lesson we will see why sometimes it is helpful to write 0.3 as 0.300.

example 47.1   Compare: 0.3 ◯ 3.0

solution   The place values are not the same. Three ones is greater than three tenths. 0.3 < 3.0

practice Compare:

a. 4.0 ◯ 4.00 $=$

b. 0.3 ◯ 0.300 $=$

c. 0.1 ◯ 0.1 $=$

d. 0.6 ◯ 0.600 $-$

e. 0.2 ◯ 0.02 $>$

f. 0.03 ◯ 0.30 $<$

g. 0.12 ◯ 0.120 $=$

h. 1.50 ◯ 1.500 $=$

problem set 47 Draw the decimals chart.

1. The average of two numbers is 16. If one of the numbers is 11, what is the other number? 21

2. Sixty miles per hour is the same speed as how many miles per minute? 1 mpm

3. The movie started at 11:45 A.M. and ended at 1:20 P.M. How many hours and minutes did it last? 1 hr. 35 min.

*4. Compare: 0.3 ◯ 3.0 $<$

5. Compare: 0.12 ◯ 0.120 $=$

6. $0.1 + 0.2 + 0.3 + 0.4 =$ 1

7. $8 \times 0.125 =$ 1

8. $3 - 2.1 =$ 0.9

9. Estimate the sum of 4967, 8142, and 6890. 20,000

10. $0.8 \overline{)0.144}$ 0.18

11. $0.6 \overline{)12}$ 20

12. $8 \overline{)0.9}$ 0.1125

13. What is the price of 100 pens at 39¢ each? $39

14. Write $(5 \times 10) + (6 \times \frac{1}{10}) + (4 \times \frac{1}{100})$ as a standard numeral. 50.64

15. What is the least common multiple of 6 and 8? 24

16. $5\frac{5}{12} + 7\frac{7}{12} =$ 13

17. $12 - 5\frac{2}{3} = 6\frac{1}{3}$

18. $5\frac{1}{4} - 2\frac{3}{4} = 2\frac{1}{2}$

19. What is $\frac{5}{6}$ of 60? 50

20. $\frac{9}{10} \div \frac{2}{3} = 1\frac{7}{20}$

21. The temperature rose from $-12°$ to $5°$. How many degrees did the temperature rise? 17°

22. What number is halfway between 463 and 643?   553

23. $\dfrac{3}{8} = \dfrac{?}{24}$ 9

24. One square foot is how many square inches?   144 sq. in.

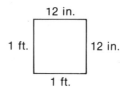

25. How long is the line segment? $3\frac{3}{8}$ in.

LESSON 48

Comparing Decimal Fractions

Decimal fractions can be compared easily if the decimal numbers being compared have the **same number of decimal places**. Comparing 0.300 and 0.123, we can easily see that 0.300 is larger. When comparing decimal numbers which do not have the same number of decimal places, we should fill places with zeros so that the numbers do have the same number of decimal places. When comparing 0.3 and 0.123 we will avoid the careless error of thinking that 0.123 is larger if we fill places after 0.3 so that it has the same number of decimal places as the number to which it is being compared. 0.300 > 0.123

example 48.1   Compare: 0.2 ◯ 0.12

solution   **First fill places with zero** so that the numbers being compared have the same number of places **after the decimal point**; then compare: 0.20 > 0.12

practice First fill places with zeros so that the numbers being compared have the same number of decimal places; then compare.

a. 0.1 ◯ 0.01 >   **f.** 4.25 ◯ 0.425 >
b. 0.2 ◯ 0.15 >   **g.** 0.012 ◯ 0.12 <
c. 1.2 ◯ 1.23 <   **h.** 0.001 ◯ 0.1 <
d. 12.5 ◯ 1.25 >   **i.** 0.31 ◯ 0.039 >
e. 0.63 ◯ 0.625 >   **j.** 0.4 ◯ 0.40 =

problem set 48 Draw the decimals chart.

1. When the product of 0.2 and 0.3 is subtracted from the sum of 0.2 and 0.3, what is the difference? 0.44

2. Colorado is a rectangular shape. It is about 384 miles east to west and 273 miles north to south. It covers an area of about how many square miles? (Estimate.) 104,832 or 120,000 sq. mi.

3. Dolores went to sleep at 9:15 P.M. and woke up at 7:15 A.M.. How long did she sleep? 10 hr.

*4. Compare: 0.2 ◯ 0.12 >   5. Compare: 0.3 ◯ 0.300 =

6. 0.67 + 2 + 1.33 = 4   7. 12 × 0.25 = 3

8. 0.07 ⟌ 3.5   50   9. 0.5 ⟌ 12   24   10. 8 ⟌ 0.14   0.0175

11. $15 - 7\frac{1}{4} = 7\frac{3}{4}$ 12. $6\frac{1}{8} - 4\frac{3}{8} = 1\frac{3}{4}$ 13. $\frac{5}{6} = \frac{?}{24}$ 20

14. 5 − 1.37 = 3.63   15. 0.012 × 1.5 = 0.018

16. Write the decimal numeral one and twelve thousandths. 1.012

17. $5\frac{7}{10} + 4\frac{9}{10} = 10\frac{3}{5}$ 18. $\frac{5}{2} \times \frac{5}{3} = 4\frac{1}{6}$ 19. $\frac{9}{10} \div \frac{3}{4} = 1\frac{1}{5}$

20. There are 24 hours in a day. Jim sleeps 8 hours each night. Eight hours is what fraction of a day? $\frac{1}{3}$

21. List the factors which 12 and 18 have **in common**. (List the numbers which are factors of **both**.) 1, 2, 3, 6

22. What is the average of 1.2, 1.3, and 1.7?   1.4

23. Estimate the difference of 5670 and 3940 to the nearest thousand.   2000

24. $563 - 635 = $ -72

25. How long is the pencil eraser? $2\frac{1}{8}$ in.

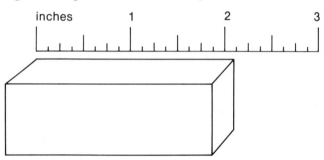

LESSON
49

Decimal Number Line (Tenths)

We can locate different kinds of numbers on the number line. We have learned to locate whole numbers, negative numbers, and fractions on the number line. Since decimal fractions are a kind of fraction, we can also locate decimal numbers on the number line.

The distance between the whole numbers has been divided into 10 equal lengths. Each length is $\frac{1}{10}$ of the distance between the whole numbers. The arrow is pointing to a mark three spaces beyond the 1. The mark it is pointing to is $1\frac{3}{10}$. We can rename $\frac{3}{10}$ as the decimal 0.3 so we can say that the arrow is pointing to the mark 1.3. When a unit has been divided into 10 spaces we normally use the decimal form instead of the fractional form to name the mark.

practice   To which **decimal number** is each arrow pointing?

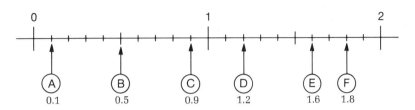

problem   Draw the decimals chart.
set 49

1. The first three positive odd numbers are 1, 3, and 5. What
    is the sum of the first 10 positive odd numbers?   100

2. One hundred centimeters equals one meter. A flagpole
    1400 centimeters tall is how many meters tall?   14 m

3. The boxing match ended after two minutes of the 12th
    round. Each round lasts three minutes. For how many
    minutes did the contenders box?   35 min.

4. Compare: 3.4 ◯ 3.389   >   5. Compare: 0.60 ◯ 0.600   =

6. 7.25 + 2 + 0.75   10   7. 3.75 × 2.4 = 9

8. 1 − 0.97 =   9. 0.12)7.2   10. 0.4)7   11. 6)0.138
    0.03   60   17.5   0.023

12. $1 - \dfrac{5}{12} = \dfrac{7}{12}$ 13. $6\dfrac{1}{8} - 1\dfrac{7}{8} = 4\dfrac{1}{4}$ 14. $\dfrac{3}{4} = \dfrac{?}{24}$ 18

15. Write 0.07 as a fraction. $\dfrac{7}{100}$

16. Which digit in 4.637 is in the same place as the 2 in 85.21?
     6

17. How many square centimeters equal one square meter?
     10,000 sq. cm

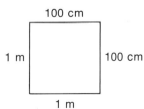

18. What is the least common multiple of 6 and 9?   18

19. $6\dfrac{5}{8} + 4\dfrac{5}{8} = 11\dfrac{1}{4}$ **20.** $\dfrac{8}{3} \times \dfrac{3}{1} = 8$ **21.** $\dfrac{8}{3} \div \dfrac{3}{1} = \dfrac{8}{9}$

22. If you sleep 8 hours each day, what fraction of the day do you **not** sleep? $\frac{2}{3}$

23. Find the average of 2.4, 6.3, and 5.7.   4.8

24. Which factors do 18 and 24 have in common?   1, 2, 3, 6

25. To which decimal number is the arrow pointing?   3.9

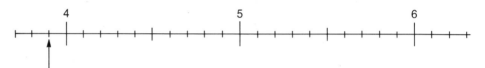

LESSON 50

Rounding Decimal Numbers

It is often necessary to round decimal numbers. Money, for instance, is usually rounded to two places after the decimal point because we do not have a coin smaller than one hundredth of a dollar. In this lesson we will learn a three-step method for rounding decimal numbers. We find that it helps to use a circle and an arrow.

1. Mark the place to which the number is to be rounded with a circle and draw an arrow over the following place.
2. Drop the arrow-marked digit and all digits to its right.
3. If the arrow-marked digit is 5 or more, increase the circled digit by 1. If the arrow-marked digit is less than 5, leave the circled digit unchanged.

example 50.1   Round $0.1625 to the nearest cent.

solution We are trying to find out whether $0.1625 is closer to $0.16 or $0.17. Following our three steps, we first mark the cents' place (the second place after the decimal point) with a circle and the following place with an arrow.

Next we drop the arrow-marked digit and all digits to its right. Finally we notice that the arrow-marked digit is less than 5, so we leave the circled digit unchanged. Answer: **$0.16**

$ 0 . 1 ⑥ \overset{\downarrow}{2} 5$

$ 0 . 1 ⑥$

practice
(See Practice
Set S in the
Appendix.)

Round to the tenths' place.

a. 0.12   0.1
b. 1.23   1.2
c. 4.56   4.6
d. 12.345   12.3
e. 0.446   0.4

Round to the hundredths' place.

f. 0.123   0.12
g. 0.456   0.46
h. 1.2854   1.29
i. $3.333 $3.33
j. $6.6666 $6.67

problem set 50 Draw the decimals chart.

1. When the third multiple of 8 is subtracted from the fourth multiple of 6, what is the difference?   0

2. Traveling 60 miles per hour, how long would it take a train to travel 240 miles from Dallas to Houston?   4 hr.

3. Napoleon I was born in 1769. How old was he when he was crowned emperor of France in 1804?   35 years

*4. Round $0.1625 to the nearest cent. $0.16

5. Round 2.375 to the nearest tenth.   2.4

6. Compare: 0.4 ◯ 0.399   >

7. $0.125 + 0.25 + 0.375 = 0.75$ 8. $0.4 - 0.399 = 0.001$

9. $0.25\overline{)4}$ 16   10. $8\overline{)0.3}$ 0.0375   11. $20 - 17\dfrac{3}{4} = 2\frac{1}{4}$

12. $3\frac{5}{12} - 1\frac{7}{12} = 1\frac{5}{6}$ **13.** $\frac{5}{8} = \frac{?}{24}$ 15 **14.** $\frac{4}{5} \times \square = 1$ $\frac{5}{4}$

15. $0.19 \times 0.21 = 0.0399$ **16.** Write 0.01 as a fraction. $\frac{1}{100}$

17. Write $(6 \times 10) + (7 \times \frac{1}{100})$ as a decimal numeral. 60.07

18. How many square inches are needed to cover the area of the rectangle? 288 sq. in.

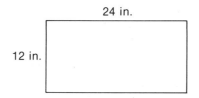

19. Which is the least common multiple of 2, 3, and 4? 12

20. $5\frac{3}{10} + 6\frac{9}{10} = 12\frac{1}{5}$ **21.** $\frac{10}{3} \times \frac{1}{2} = 1\frac{2}{3}$ **22.** $\frac{10}{3} \div \frac{1}{2} = 6\frac{2}{3}$

23. Estimate the quotient when 4876 is divided by 98. 50

24. Which factors do 16 and 24 have in common? 1, 2, 4, 8

25. To which decimal number is the arrow pointing? 1.5

LESSON 51

Multiplying and Dividing by 10 or 100

When you multiply by 10 or 100 you probably just "add zeros" to the number you multiplied. ($5 \times 10 = 5\underline{0}$, $5 \times 100 = 5\underline{00}$) You may think you are "adding zeros," but you are actually changing the place value of the digits in the number you are multiplying.

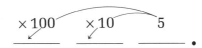

$$\times 100 \qquad \times 10 \qquad 5 \qquad \text{.}$$

This 5 is in the ones' place. When we multiply by 10, it moves over one place, and when we multiply by 100, it moves over two places. It is important to know this because when you multiply decimal numbers by 10 or 100 you will not get the right answer if you just "add zeros." When multiplying or dividing by 10 or 100 the digits actually change place values. When multiplying or dividing by 10, the digits shift one place. When multiplying or dividing by 100, the digits shift two places. Although it is actually the digits that are shifting places, we get the same effect by shifting the decimal point one place or two.

example 51.1 $3.75 \times 10 =$

solution   We could follow the chart rule "multiply, then count." Instead, we will just shift the decimal point one place to the right.

$$3.75 \times 10 \quad \longrightarrow \quad \text{means shift 1 place} \quad \longrightarrow \quad \mathbf{37.5}$$

one place

example 51.2 $3.75 \times 100 =$

solution $3.75 \times 100 \quad \longrightarrow \quad \text{means shift 2 places} \quad \longrightarrow \quad \mathbf{375.}$

two places

example 51.3 $3.75 \div 10 =$

solution   In this division we shift the decimal point to the left. To divide by 10, we shift one place; to divide by 100, we shift two places, etc.

$$3.75 \div 10 \quad \longrightarrow \quad \text{means shift one place} \quad \longrightarrow \quad \mathbf{0.375}$$

one place

practice
(See Practice
Set T in the
Appendix.)

a. $4.2 \times 10 = 42$

b. $4.2 \times 100 = 420$

c. $0.35 \times 10 = 3.5$

d. $0.35 \times 100 = 35$

e. $0.125 \times 10 = 1.25$

f. $37.5 \div 10 = 3.75$

g. $37.5 \div 100 = 0.375$

h. $1.2 \div 10 = 0.12$

i. $1.2 \div 100 = 0.012$

j. $0.12 \div 10 = 0.012$

**problem
set 51** Draw the decimals chart.

1. What is the product of one half and two thirds? $\frac{1}{3}$

2. A piano has 88 keys. Fifty-two of the keys are white. How many more white keys are there than black keys? 16

3. The deepest part of the Atlantic Ocean is thirty thousand, two hundred forty-six feet. Write that number. 30,246

*4. $3.75 \times 10 =$ 37.5   *5. $3.75 \div 10 =$ 0.375

6. $2 \times 2 \times 2 \times 2 \times 2 =$ 32   7. Convert and reduce: $\dfrac{150}{12}$
                                                              $12\frac{1}{2}$

8. Multiply and simplify: $0.125 \times 4 =$ 0.5

9. $(1 + 0.2) \div (1 - 0.2) =$ 1.5   10. $5\dfrac{1}{3} - 1\dfrac{2}{3} =$ $3\dfrac{2}{3}$

11. $\dfrac{5}{2} \times \dfrac{4}{1} =$ 10   12. $\dfrac{3}{5} \div \dfrac{2}{3} =$ $\dfrac{9}{10}$ 13. $\$10 - \$0.10 =$
                                                                                              $9.90

14. Which digit in 6.789 is in the hundredths' place? 8

15. Round 12.475 to the nearest tenth. 12.5

16. Which is the least amount?   1.02   1.20   0.21   0.21

17. What number is missing in this sequence? 1, 2, 4, 7, 11, ___, 22 16

18. The perimeter of a square room is 80 feet. How many floor tiles 1 foot square would be needed to cover the area of the room? 400 sq. ft.

19. What fraction of a foot is 3 inches? $\frac{1}{4}$

20. $\square - 20 = 5$ 25   21. List the factors of 17. 1, 17

22. What is the least common multiple of 2, 4, and 6? 12

23. $\dfrac{4}{4} - \dfrac{2}{2} = 0$

24. A meter is about one **big** step. About how many meters above the floor is the top of the chalkboard?   probably 2

25. To which decimal mark is the arrow pointing?   1.8

LESSON
52

Forms of Money: Dollars, Cents

There are two different ways of writing amounts of money, either (1) as a number of cents, or (2) as a number of dollars. You can never write it both ways at the same time.

Amounts less than a dollar are sometimes written as a number of cents. A dime is ten cents and can be written 10¢.

Amounts less than a dollar can also be written as a fraction of a dollar. A dime is $\frac{10}{100}$ of a dollar. Ten-hundredths can be written as a decimal, 0.10, so a dime can be written as $0.10.

When the two different forms of expressing money are in the same problem, one must be changed so that all amounts are in the same form before the problem can be solved.

example 52.1 $2 − 23¢ =

solution   The cent form must be changed to dollar form, $0.23, before subtracting.

$$\begin{array}{r} \$2.00 \\ -\ \$0.23 \\ \hline \mathbf{\$1.77} \end{array}$$

practice   Change to dollar form.

a. 23¢ $0.23
b. 5¢ $0.05
c. 1¢ $0.01
d. 99¢ $0.99

Solve and write answer in the indicated form.

e. 63¢ + 45¢ = $1.08
f. $1 − 17¢ = $0.83
g. $1 − 71¢ = 29¢
h. $0.12 + 8¢ = 20¢

problem   Draw the decimals chart.
set 52

1. What is the difference between the sum of 0.2 and 0.3 and the product of 0.2 and 0.3?   0.44

2. The turkey must cook for 4 hours and 45 minutes. What time must it be put in the oven in order to be done by 3:00 P.M.?   10:15 A.M.

3. Billy won the contest by eating $\frac{1}{4}$ of a berry pie in 7 seconds. At this rate how long would it take Billy to eat a whole berry pie?   28 sec.

*4. $2 − 23¢ = $1.77   5. $0.96 + 8¢ = $1.04

6. What is the average of 47, 52, 63, and 66?   57

7. $3.6 + 4 + 0.39 = 7.99$ 8. $0.375 \times 100 = 37.5$

9. $36 \div 0.12 = 300$ 10. $0.15 \div 4 = 0.0375$

11. $6\frac{1}{4} - 3\frac{3}{4} = 2\frac{1}{2}$ 12. $\frac{2}{3} \times \frac{3}{5} = \frac{2}{5}$ 13. $\frac{2}{3} \div \frac{3}{5} = 1\frac{1}{9}$

14. Which digit in 3,456 has the same place value as the 2 in 28.7?   5

15. Round 0.416 to the nearest hundredth.   0.42

16. Which is closest to 1? 1.2, 0.9, 0.1   0.9

17. What is the sum of the first eight positive odd numbers?   64

18. What is the perimeter of the square? $1\frac{1}{2}$ in. $\boxed{}$ $\frac{3}{8}$ in.

19. What fraction of a yard is 3 inches? $\frac{1}{12}$

20. $8 \times \boxed{} = 1000$ 125   21. $12 − 123 = −111$

22. List the factors of 26.   1, 2, 13, 26

23. What is the smallest number which is a multiple of both 6 and 9?   18

24. How long is the line segment? $3\frac{3}{8}$ in.

25. Here is part of the multiplication table. What number is missing? 49

36	42	48
42	?	56
48	56	64

LESSON
53

Fractions Chart

In Lesson 36 we discussed the decimals chart. The decimals chart helps us remember the six rules for decimal numbers. Now we will look at the fractions chart. This chart tells us to use the letters S.O.S. to help with fractions. These letters can help us remember the three steps for dealing with fractions:

Shape—Operate—Simplify.

Fractions Chart

	+ −	**× ÷**	
SHAPE	common denominators	fraction form	
		×	÷
OPERATE	+ − top only	$\dfrac{t \times t}{b \times b}$	invert 2nd and multiply
		cancel	
SIMPLIFY	reduce/convert		

We are already familiar with the Operate and Simplify rules from earlier lessons. The Shape rules need more explanation.

We can only add or subtract "like" things. "Like" fractions have the same bottom number—**common denominators**. The fractions in $\frac{1}{2} + \frac{1}{3}$ do not have common denominators but the fractions in $\frac{3}{6} + \frac{2}{6}$ do.

When we multiply or divide fractions the fractions must be in **fraction form**—that means no whole numbers or mixed numbers. $1\frac{2}{3} \times 2$ is not in fraction form; $\frac{5}{3} \times \frac{2}{1}$ is.

In the lessons which follow we will explain how to write fraction problems in the right "shape." Meanwhile, we ask you to memorize the chart.

problem set 53

Draw the fractions chart.

1. If 0.4 is the dividend and 4 is the divisor, what is the quotient?   0.1

2. In 1900 the U.S. population was 76,212,168. In 1950 the population was 151,325,798. **Estimate** the increase in population between 1900 and 1950 to the nearest million.   75,000,000

3. Mark was $59\frac{3}{4}$ inches tall when he turned 11 and $61\frac{1}{4}$ inches tall when he turned 12. How many inches did he grow during the year? $1\frac{1}{2}$ inches

4. $1000 - (100 - 1) = 901$

5. $\dfrac{1000}{24} = 41\frac{2}{3}$

6. What number is 25 less than 7? -18

7. What number is halfway between 37 and 143?   90

8. $\$3 - 24¢ = \$2.76$

9. $(1.2 \div 0.12) - 1.2 = 8.8$

10. $4.2 \div 100 = 0.042$

11. $6\frac{2}{5} - 3\frac{4}{5} = 2\frac{3}{5}$

12. $\frac{4}{3} \times \frac{4}{3} = 1\frac{7}{9}$

13. $\frac{4}{3} \div \frac{4}{3} = 1$

14. Which digit is in the hundred-thousands' place in 123,456,789?   4

15. Round $26.777 to the nearest cent. $26.78

16. Compare: $123 - 45 \bigcirc 123 - 46$ >

17. What is the twelfth number in this sequence? 1, 3, 5, 7, 9, . . .
23

18. How many square feet of tile would be needed to cover the area of a room 14 feet long and 12 feet wide?
168 sq. ft.

19. Nine of the 30 students received A's on the test. What fraction of the students received A's? $\frac{3}{10}$

20. $\frac{5}{6} = \frac{?}{12}$ 10

21. What is the least common multiple of 3, 4, and 6?   12

22. What is the reciprocal of $\frac{9}{10}$? $\frac{10}{9}$

23. Write 0.7 as a fraction. $\frac{7}{10}$

24. Which is longer—1 cm or 1 in.?   1 in.

25. A line 100 inches long would be how many centimeters long?   254 cm

Conversion Table

METRIC UNITS	CUSTOMARY UNITS
1 cm	0.394 in.
2.54 cm	1 in.

LESSON 54

Common Denominators, Part 1

When we add fractions we are counting how many **equal-size** parts we have in the total. The denominator of a fraction tells the size of each part, so the denominators must be the same before the fractions can be added.

In the illustration at right, we cannot count the shaded parts to name the fraction answer because the sizes of the parts are different. However, if we cut the first circle into fourths we will have equal parts which we **can** add: $\frac{2}{4} + \frac{1}{4} = \frac{3}{4}$.

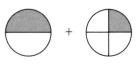

Remember, the name of a fraction can be changed by multiplying the fraction by a different name for 1. The fraction $\frac{1}{2}$ can be changed to fourths by multiplying by $\frac{2}{2} \cdot \frac{1}{2} \times \frac{2}{2} = \frac{2}{4}$. This is the method we use to add or subtract fractions with unlike denominators.

example 54.1 $\dfrac{1}{4} + \dfrac{1}{2} =$

solution The names of the fractions must be changed so that the fractions have common denominators. Multiply $\frac{1}{2} \times \frac{2}{2}$ so that both fractions are fourths; then add.

$$\frac{1}{4} \times \frac{1}{1} = \frac{1}{4}$$
$$+\frac{1}{2} \times \frac{2}{2} = \frac{2}{4}$$
$$\overline{\qquad \frac{3}{4}}$$

example 54.2

$$\frac{1}{2}$$
$$-\frac{1}{8}$$
$$\overline{\qquad}$$

solution

$$\frac{1}{2} \times \frac{4}{4} = \frac{4}{8}$$
$$-\frac{1}{8} \times \frac{1}{1} = \frac{1}{8}$$
$$\overline{\qquad \frac{3}{8}}$$

practice

a. $\begin{array}{r} \frac{1}{2} \\ +\frac{3}{8} \\ \hline \frac{7}{8} \end{array}$ b. $\begin{array}{r} \frac{3}{8} \\ +\frac{1}{4} \\ \hline \frac{5}{8} \end{array}$ c. $\begin{array}{r} \frac{3}{4} \\ +\frac{1}{8} \\ \hline \frac{7}{8} \end{array}$ d. $\begin{array}{r} \frac{2}{3} \\ +\frac{1}{9} \\ \hline \frac{7}{9} \end{array}$

e. $\begin{array}{r} \frac{1}{2} \\ -\frac{1}{4} \\ \hline \frac{1}{4} \end{array}$ f. $\begin{array}{r} \frac{5}{8} \\ -\frac{1}{4} \\ \hline \frac{3}{8} \end{array}$ g. $\begin{array}{r} \frac{3}{4} \\ -\frac{3}{8} \\ \hline \frac{3}{8} \end{array}$

problem set 54

Draw the fractions chart.

1. What is the sum of twenty-one thousand, five hundred and ten thousand, nine hundred fifty?   32,450

2. The Babylonian Empire was at its greatest power between 612 B.C. and 539 B.C. How many years was that?   73 years

3. The average pumpkin weighs 6 pounds. The Great Pumpkin weighs 324 pounds. The Great Pumpkin weighs as much as how many average pumpkins?   54

*4. $\dfrac{1}{4} + \dfrac{1}{2} = \dfrac{3}{4}$

*5. $\begin{array}{r} \frac{1}{2} \\ -\frac{1}{8} \\ \hline \frac{3}{8} \end{array}$

6. $\begin{array}{r} \frac{1}{2} \\ -\frac{1}{6} \\ \hline \frac{1}{3} \end{array}$

7. $6.28 + 4 + 0.13 = $ 10.41

8. $81 \div 0.9 = $ 90

9. $0.2 \div 10 = $ 0.02

10. $0.17 \times 100 = $ 17

11. $3\dfrac{1}{4} - \dfrac{3}{4} = 2\dfrac{1}{2}$

12. $\dfrac{5}{6} \times \dfrac{2}{3} = \dfrac{5}{9}$

13. $\dfrac{9}{4} \div \dfrac{2}{3} = 3\dfrac{3}{8}$

14. Write in standard notation: $(6 \times 10{,}000) + (4 \times 100) + (2 \times 10)$ 60,420

15. Multiply 0.14 and 0.8 and round the product to the nearest hundredth.   0.11

16. Compare: $\dfrac{2}{3} \bigcirc \dfrac{2}{3} \times \dfrac{2}{2}$ =

17. Which is closest to 1? $\dfrac{1}{4}, \dfrac{1}{2}, \dfrac{3}{4}$ $\dfrac{3}{4}$

18. A 20-foot rope was used to make a square. How many square feet of area are enclosed by the rope?   25 sq. ft.

19. What fraction of a dollar is four dimes? $\frac{2}{5}$

20. $6 - 3 = 7 - \square$ 4

21. List the factors of 23.
1, 23

22. How many 12's are in 1212?   101

23. $\dfrac{2}{5} \times \square = 1$ $\dfrac{5}{2}$

24. How long is the line segment?
$\frac{7}{8}$ in.

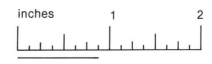

25. Which of these is a cube?   (b)

(a)   (b)   (c)

LESSON
55

Common Denominators, Part 2

The fractions $\frac{1}{2}$ and $\frac{1}{3}$ cannot be added in their present form. They have different sized parts.

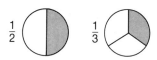

Dividing the first circle into fourths will not help. The parts still have different sizes.

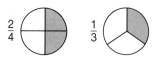

Both fractions must have the same size parts if they are to be added.

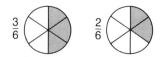

Often both fractions of an addition or subtraction problem must be renamed. The new names should have denominators equal to the least common multiple of the two denominators.

example 55.1 $\frac{1}{2}$
$+\frac{1}{3}$

solution The denominators are 2 and 3. The LCM of 2 and 3 is 6. Rename the fractions so that 6 is the denominator in each. Then add.

$$\frac{1}{2} \times \frac{3}{3} = \frac{3}{6}$$
$$+\frac{1}{3} \times \frac{2}{2} = \frac{2}{6}$$
$$\overline{\frac{5}{6}}$$

example 55.2 $\frac{3}{4}$
$-\frac{2}{3}$

solution The least common multiple of 4 and 3 is 12. Rename both fractions so that their denominators are 12, then subtract.

$$\frac{3}{4} \times \frac{3}{3} = \frac{9}{12}$$
$$-\frac{2}{3} \times \frac{4}{4} = \frac{8}{12}$$
$$\overline{\frac{1}{12}}$$

practice **a.** $\frac{2}{3}$ **b.** $\frac{1}{4}$ **c.** $\frac{3}{4}$ **d.** $\frac{2}{3}$ **e.** $\frac{1}{3}$

$-\frac{1}{2}$ $+\frac{1}{3}$ $-\frac{1}{3}$ $+\frac{1}{4}$ $-\frac{1}{4}$

$\frac{1}{6}$ $\frac{7}{12}$ $\frac{5}{12}$ $\frac{11}{12}$ $\frac{1}{12}$

problem set 55   Draw the fractions chart.

1. How many 4's are in 88?   22

2. Of the 88 keys on a piano, 52 are white. What fraction of a piano's keys are white? $\frac{13}{22}$

3. If $4\frac{1}{2}$ apples are needed to make an apple pie, how many apples would be needed to make four apple pies?   18

*4. $\frac{1}{2}$ ***5.** $\frac{3}{4}$ **6.** $\frac{2}{3}$ **7.** $\frac{1}{3}$

$+\frac{1}{3}$ $-\frac{2}{3}$ $-\frac{1}{2}$ $-\frac{1}{4}$

$\frac{5}{6}$ $\frac{1}{12}$ $\frac{1}{6}$ $\frac{1}{12}$

8. $\$3 + \$1.75 + 65¢ = $ \$5.40   **9.** $0.625 \times 0.4 = $ 0.25

10. $24 \div 0.08 = $ 300   **11.** $3\frac{1}{8} - 1\frac{7}{8} = $ $1\frac{1}{4}$

12. $\dfrac{5}{8} \times \dfrac{2}{3} = $ $\dfrac{5}{12}$ **13.** $\dfrac{5}{8} \div \dfrac{2}{3} = $ $\dfrac{15}{16}$

14. Write as a decimal numeral: $(8 \times 10) + (6 \times \frac{1}{10}) + (5 \times \frac{1}{100})$
80.65

15. Estimate the sum of 3627 and 4187 to the nearest hundred.
7800

16. Which is between 1 and 2? 1.875, 2.01, 0.15   1.875

17. What is the average of 1.2, 1.3, 1.4, and 1.5?   1.35

18. The area of a square is 36 square inches. What is its perimeter?   24 in.

19. Returning from the store, Mom found that four of the dozen eggs were cracked. What fraction of the eggs were cracked?
$\frac{1}{3}$

20. $\square \times 1 = \dfrac{2}{3}$ $\dfrac{2}{3}$ **21.** $\square \times \dfrac{2}{3} = 1$ $\dfrac{3}{2}$ **22.** $\dfrac{2}{3} - \square = 0$ $\dfrac{2}{3}$

23. What is the smallest number which is a multiple of both 8 and 10?   40

Use this chart of test results to answer questions 24 and 25.

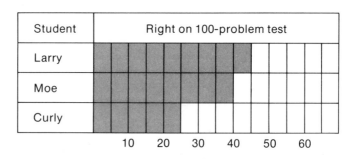

24. How many problems did Moe miss on the test?   60

25. How many more problems did Larry have right on the test than Curly?   20

LESSON
56

Adding and Subtracting Fractions—
Three Steps

The three steps for solving fraction problems are:

1. Put the problem in the right **shape** to do the operation.
2. Do what the **operation** sign indicates ($+$, $-$, $\times$, $\div$).
3. **Simplify** the answer (reduce, convert).

When adding and subtracting fractions, the steps are:

1. Shape—write fractions with common denominators.
2. Operate—add or subtract numerators only.
3. Simplify—reduce or convert where possible.

example 56.1 $\dfrac{1}{2} + \dfrac{2}{3} =$

solution   Write the problem with common denominators. Add the numerators. Simplify the answer. In this case, convert.

$$\begin{array}{r} \frac{1}{2} \times \frac{3}{3} = \frac{3}{6} \\ + \frac{2}{3} \times \frac{2}{2} = \frac{4}{6} \\ \hline \frac{7}{6} = 1\frac{1}{6} \end{array}$$

example 56.2 $\dfrac{1}{2} + \dfrac{1}{6} =$

solution   Write the problem with common denominators. Add the numerators. Simplify the answer. Here we can reduce.

$$\begin{array}{r} \frac{1}{2} \times \frac{3}{3} = \frac{3}{6} \\ + \frac{1}{6} \times \frac{1}{1} = \frac{1}{6} \\ \hline \frac{4}{6} = \frac{2}{3} \end{array}$$

example 56.3 $\dfrac{5}{6} + \dfrac{2}{3} =$

solution   Write the problem with common denominators. Add the numerators. Simplify. Here we can convert **and** reduce.

$$\begin{array}{r} \frac{5}{6} \times \frac{1}{1} = \frac{5}{6} \\ + \frac{2}{3} \times \frac{2}{2} = \frac{4}{6} \\ \hline \frac{9}{6} = 1\frac{3}{6} \\ = 1\frac{1}{2} \end{array}$$

practice
(See Practice Set V in the Appendix.)

a. $\dfrac{1}{2} + \dfrac{1}{6} = \dfrac{2}{3}$ e. $\dfrac{1}{2} + \dfrac{5}{6} = 1\dfrac{1}{3}$ i. $\dfrac{7}{10} - \dfrac{1}{2} = \dfrac{1}{5}$

b. $\dfrac{1}{2} + \dfrac{2}{3} = 1\dfrac{1}{6}$ f. $\dfrac{2}{3} - \dfrac{1}{6} = \dfrac{1}{2}$ j. $\dfrac{7}{10} - \dfrac{1}{5} = \dfrac{1}{2}$

c. $\dfrac{2}{3} + \dfrac{3}{4} = 1\dfrac{5}{12}$ g. $\dfrac{5}{6} - \dfrac{1}{2} = \dfrac{1}{3}$

d. $\dfrac{5}{6} + \dfrac{5}{12} = 1\dfrac{1}{4}$ h. $\dfrac{1}{2} - \dfrac{1}{6} = \dfrac{1}{3}$

problem set 56   Draw the fractions chart.

1. If $\frac{1}{2}$ is the dividend and $\frac{1}{3}$ is the divisor, then what is the quotient? $1\frac{1}{2}$

2. Thomas Jefferson was born in 1743. How old was he when he became president of the United States in 1801?
58 years

3. The smallest two-digit number is 10. The largest two-digit number is 99. Altogether, how many different two-digit numbers are there?   90

*4. $\frac{1}{2} + \frac{2}{3} = 1\frac{1}{6}$ *5. $\frac{1}{2} + \frac{1}{6} = \frac{2}{3}$ *6. $\frac{5}{6} + \frac{2}{3} = 1\frac{1}{2}$

7. How many years was it from 4 B.C. to 29 A.D.?   32 years

8. $\$32.50 \div 10 = \$3.25$ 9. $2 - (1 - 0.2) = 1.2$

10. $6 \div 0.12 = 50$ 11. $5\frac{3}{8} - 2\frac{5}{8} = 2\frac{3}{4}$

12. $\frac{3}{4} \times \frac{3}{5} = \frac{9}{20}$ 13. $\frac{3}{4} \div \frac{3}{5} = 1\frac{1}{4}$

14. Name the place value of the 7 in 3.567.   thousandths

15. Divide 0.5 by 4 and round the quotient to the nearest tenth.
0.1

16. Rearrange these in order of size from least to greatest: 0.3, 3.0, 0.03   0.03, 0.3, 3.0

17. What is the twentieth number in this sequence? 2, 4, 6, 8, ...   40

18. What is the perimeter of the rectangle?
$1\frac{1}{2}$ in.

$\frac{1}{4}$ in.

$\frac{1}{2}$ in.

19. What is $\frac{5}{8}$ of 80?   50   20. $12 - \square = -2$ 14

21. List the factors of 29.   1, 29

22. What is the least common multiple of 12 and 18?   36

23. $\frac{3}{4} \div \square = 1\frac{3}{4}$

24. Which temperature is shown on the thermometer?
−4°

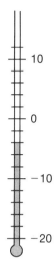

25. Which names this shape?   cylinder
cylinder   sphere   cone   cube

LESSON
57

Comparing Fractions—
Common Denominators

It is easy to compare fractions which have common denominators. We simply compare the numerators.

Compare: $\dfrac{4}{6} \bigcirc \dfrac{5}{6}$

Obviously, $\frac{5}{6}$ is greater. It is not as easy to compare fractions which do not have common denominators.

Compare: $\dfrac{3}{8} \bigcirc \dfrac{1}{2}$

One way to compare fractions which do not have common denominators is to rename one or both fractions so that they do have common denominators.

example 57.1   Compare: $\dfrac{3}{8} \bigcirc \dfrac{1}{2}$

solution We can rename $\frac{1}{2}$ as eighths by multiplying by the form of one which is $\frac{4}{4}$. So $\frac{1}{2} \times \frac{4}{4} = \frac{4}{8}$. Now we can compare:

$$\frac{3}{8} < \frac{4}{8}$$

example 57.2 Compare: $\dfrac{2}{3} \bigcirc \dfrac{3}{4}$

solution We must rename both fractions. The common denominator is the least common multiple of 3 and 4 which is 12. $\frac{2}{3} \times \frac{4}{4} = \frac{8}{12}$, $\frac{3}{4} \times \frac{3}{3} = \frac{9}{12}$. Now we can compare:

$$\frac{8}{12} < \frac{9}{12}$$

practice Write with common denominators and compare.

a. $\dfrac{1}{2} \bigcirc \dfrac{1}{4}$ $\dfrac{2}{4} > \dfrac{1}{4}$ c. $\dfrac{2}{3} \bigcirc \dfrac{1}{2}$ $\dfrac{4}{6} > \dfrac{3}{6}$ e. $\dfrac{5}{6} \bigcirc \dfrac{3}{4}$ $\dfrac{10}{12} > \dfrac{9}{12}$

b. $\dfrac{1}{3} \bigcirc \dfrac{1}{4}$ $\dfrac{4}{12} > \dfrac{3}{12}$ d. $\dfrac{2}{5} \bigcirc \dfrac{1}{2}$ $\dfrac{4}{10} < \dfrac{5}{10}$ f. $\dfrac{2}{3} \bigcirc \dfrac{3}{5}$ $\dfrac{10}{15} > \dfrac{9}{15}$

problem set 57 Draw the fractions chart.

1. What number is $\dfrac{1}{3}$ less than $\dfrac{1}{2}$? $\dfrac{1}{6}$

2. An octave is eight consecutive musical notes. How many notes is $2\frac{1}{2}$ octaves? 20

3. In three nights Rumpelstiltskin spun $44,400 worth of gold thread. What was the average value of thread he spun each night? $14,800

*4. Compare: $\dfrac{3}{8} \bigcirc \dfrac{1}{2}$ <   *5. Compare: $\dfrac{2}{3} \bigcirc \dfrac{3}{4}$ <

6. $\dfrac{1}{2} - \dfrac{3}{8} = \dfrac{1}{8}$ 7. $\dfrac{2}{3} + \dfrac{3}{4} = 1\dfrac{5}{12}$ 8. $3 - \dfrac{5}{6} = 2\dfrac{1}{6}$

9. $32.50 \times 10 =$ $325.00   10. $6.2 \times 0.48 =$ 2.976   11. $0.9 \div 8 =$ 0.1125

12. $120 \div 0.5 = 240$ 13. $\dfrac{7}{8} \times \dfrac{8}{7} = 1$ 14. $\dfrac{5}{6} \times \square = \dfrac{10}{12}$ $\dfrac{2}{2}$

15. What number is halfway between 1.2 and 3? 2.1

16. Round 36.486 to the nearest hundredth. 36.49

17. What number is next in this sequence? 0.6, 0.7, 0.8, 0.9, ____
1 or 1.0

18. What is the area of the square?
1 sq. m

1 m
1 m

19. What fraction of two dozen is half of a dozen? $\frac{1}{4}$

20. $32 = 0.32 \times \square$ 100

21. $\frac{2}{3} \times \square = \frac{12}{18}$ $\frac{6}{6}$

22. List the factors of 27.   1, 3, 9, 27

23. How many years were there from 100 B.C. to 100 A.D.?
199 years

24. How long is the line segment?
$\frac{5}{8}$ in.

inches   1   2

25. There are two pints to a quart and four quarts to a gallon. How many pints are in ten gallons?   80 pints

LESSON 58

Adding Mixed Numbers

A number with a whole number part and a fraction part is called a **mixed number**. A mixed number is a whole number **mixed** with a fraction. We can add and subtract mixed numbers the same way we add and subtract fractions. We use the same rules.

example 58.1 $2\frac{1}{2} + 1\frac{1}{6} =$

solution   Write the fractions with common denominators. Add the whole numbers and numerators. Simplify your answer when possible.

$$2\frac{1}{2} \times \frac{3}{3} = 2\frac{3}{6}$$
$$+1\frac{1}{6} \times \frac{1}{1} = 1\frac{1}{6}$$
$$3\frac{4}{6} = 3\frac{2}{3}$$

example 58.2 $1\dfrac{1}{2} + 2\dfrac{2}{3} =$

solution   Write fractions with common denominators. Add the whole numbers and numerators. Simplify. (Here we convert.) The whole numbers are added together.

$$1\tfrac{1}{2} \times \tfrac{3}{3} = 1\tfrac{3}{6}$$
$$+\, 2\tfrac{2}{3} \times \tfrac{2}{2} = 2\tfrac{4}{6}$$
$$\overline{\phantom{+ 2\tfrac{2}{3} \times \tfrac{2}{2} = {}}3\tfrac{7}{6} = 3 + 1\tfrac{1}{6}}$$
$$= 4\tfrac{1}{6}$$

practice

a. $1\dfrac{1}{2} + 1\dfrac{1}{3} = 2\dfrac{5}{6}$ e. $3\dfrac{3}{4} + 1\dfrac{1}{3} = 5\dfrac{1}{12}$ i. $5\dfrac{1}{2} + 3\dfrac{1}{6} = 8\dfrac{2}{3}$

b. $1\dfrac{1}{2} + 1\dfrac{2}{3} = 3\dfrac{1}{6}$ f. $6\dfrac{1}{2} + 5\dfrac{1}{4} = 11\dfrac{3}{4}$ j. $2\dfrac{5}{6} + 3\dfrac{1}{2} = 6\dfrac{1}{3}$

c. $1\dfrac{1}{2} + 2\dfrac{3}{4} = 4\dfrac{1}{4}$ g. $7\dfrac{1}{2} + 4\dfrac{5}{8} = 12\dfrac{1}{8}$

d. $5\dfrac{1}{3} + 2\dfrac{1}{6} = 7\dfrac{1}{2}$ h. $1\dfrac{3}{4} + 2\dfrac{3}{8} = 4\dfrac{1}{8}$

problem set 58   Draw the fractions chart.

1. What is the product of the decimal number four-tenths and the decimal number four-hundredths?   0.016

2. Tom Thumb weighed 3 pounds. His dad weighed 180 pounds. His dad weighed how many more pounds than Tom weighed?   177 lb.

3. Pluto's greatest distance from the sun is seven billion, four hundred million kilometers. Write that number.
7,400,000,000

*4. $2\dfrac{1}{2} + 1\dfrac{1}{6} = 3\dfrac{2}{3}$ *5. $1\dfrac{1}{2} + 2\dfrac{2}{3} = 4\dfrac{1}{6}$

6. Compare: $\dfrac{1}{2} \bigcirc \dfrac{3}{5}$ <   7. Compare: $\dfrac{2}{3} \bigcirc \dfrac{6}{9}$ =

8. $8\dfrac{1}{5} - 3\dfrac{4}{5} = 4\dfrac{2}{5}$ 9. $\dfrac{3}{4} \times \dfrac{5}{2} = 1\dfrac{7}{8}$ 10. $\dfrac{2}{5} \div \dfrac{2}{4} = \dfrac{4}{5}$

11. $0.875 \times 40 =$ 35 **12.** $0.07 \div 4 =$ **13.** $30 \div 0.6 =$ 50
0.0175

14. What number is halfway between 0.1 and 0.24? 0.17

15. Round 36,428,591 to the nearest million. 36,000,000

16. What temperature is 23° less than 8°? $-15°$

17. What number is missing? 320, 160, 80, <u>40</u>, 20, 10, 5

18. How many square inches are needed to cover a square foot? 144 sq. in.

19. One centimeter is what fraction of one meter? $\frac{1}{100}$

20. $\boxed{} - 25 = 2$ 27 **21.** List the factors of 31. 1, 31

22. What is the smallest number that is a multiple of 12 and 15? 60 **23.** $32 \div 10 = 3.2 \times \boxed{}$ 1

Use the chart below to answer questions 24 and 25.

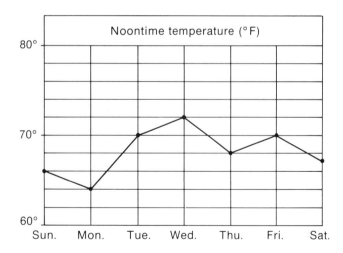

24. How many degrees difference was there between the highest and lowest noontime temperatures during the week?
8° F

25. What was the Saturday noontime temperature? 67° F

LESSON
59

Adding Three or More Fractions

Fractions can only be added when they have common denominators. When the fractions to be added do not have common denominators, we must find other names for the fractions so that they will have common denominators. We can find other names for fractions by multiplying the fractions by other names for 1, like $\frac{2}{2}$, $\frac{3}{3}$, or $\frac{4}{4}$.

Sometimes three or more fractions are to be added at the same time. We must find a common denominator for all the fractions being added. A common denominator is the least common multiple of all of the denominators. When we know what the common denominator is we can rename the fractions and add.

example 59.1 $\frac{1}{2} + \frac{1}{4} + \frac{1}{8} =$

solution   First we find the common denominator.

The LCM of 2, 4, and 8 is 8.

We rename all fractions as eighths.

Then we add and simplify if possible.

$$\frac{1}{2} \times \frac{4}{4} = \frac{4}{8}$$
$$\frac{1}{4} \times \frac{2}{2} = \frac{2}{8}$$
$$+ \frac{1}{8} \times \frac{1}{1} = \frac{1}{8}$$
$$\overline{\qquad \frac{7}{8}}$$

example 59.2 $1\frac{1}{2} + 2\frac{1}{3} + 3\frac{1}{6} =$

solution   The common denominator is 6.

We rename all fractions.

We add whole numbers and fractions.

We simplify when possible.

$$1\frac{1}{2} \times \frac{3}{3} = 1\frac{3}{6}$$
$$2\frac{1}{3} \times \frac{2}{2} = 2\frac{2}{6}$$
$$+ 3\frac{1}{6} \times \frac{1}{1} = 3\frac{1}{6}$$
$$\overline{\qquad 6\frac{6}{6} = \mathbf{7}}$$

practice   **a.** $\frac{1}{2} + \frac{3}{4} + \frac{1}{8} = \quad 1\frac{3}{8}$ **d.** $\frac{1}{2} + \frac{1}{3} + \frac{1}{6} = \quad 1$ **g.** $1\frac{1}{2} + 1\frac{1}{3} + 1\frac{1}{4} = \quad 4\frac{1}{12}$

b. $\frac{1}{2} + \frac{1}{3} + \frac{1}{4} = \quad 1\frac{1}{12}$ **e.** $\frac{1}{2} + \frac{2}{3} + \frac{5}{6} = \quad 2$ **h.** $1\frac{1}{4} + 1\frac{1}{8} + 1\frac{1}{2} = \quad 3\frac{7}{8}$

c. $\frac{1}{2} + \frac{2}{3} + \frac{1}{6} = \quad 1\frac{1}{3}$ **f.** $\frac{1}{2} + \frac{3}{4} + \frac{7}{8} = \quad 2\frac{1}{8}$ **i.** $1\frac{1}{2} + 2\frac{2}{3} + 3\frac{3}{4} = \quad 7\frac{11}{12}$

problem Draw the fractions chart.
set 59

1. What is the cost per ounce for a 42-ounce box of oatmeal priced at $1.26? $0.03

2. There are 30 days in November. How many days are there from November 19 to December 25? 36

3. The smallest three-digit number is 100. What is the largest three-digit number? 999

*4. $\frac{1}{2} + \frac{1}{4} + \frac{1}{8} = \frac{7}{8}$

*5. $1\frac{1}{2} + 2\frac{1}{3} + 3\frac{1}{6} = 7$

6. $5 - 1\frac{3}{5} = 3\frac{2}{5}$

7. $6\frac{1}{8} - 3\frac{5}{8} = 2\frac{1}{2}$

8. Compare: $\frac{5}{8} \bigcirc \frac{3}{4}$ <

9. $\frac{3}{5} \times \frac{1}{3} = \frac{1}{5}$

10. $\frac{3}{5} \div \frac{1}{3} = 1\frac{4}{5}$

11. $\frac{5}{8} \times 24 = 15$

12. $0.65 \times 0.14 = $ 0.091

13. $65 \div 0.05 = $ 1300

14. What is the place value of the 9 in 46.934? tenths

15. Round the product of 0.24 and 0.26 to the nearest hundredth. 0.06

16. What is the average of 1.3, 2, and 0.81? 1.37

17. What is the sum of the first seven numbers of this sequence? 1, 3, 5, 7, . . . 49

18. How many square feet are needed to cover a square yard? 9 sq. ft.

19. Ten centimeters is what fraction of one meter? $\frac{1}{10}$

20. $1.2 + 1.2 + 1.2 = 3 \times \boxed{}$ 1.2

21. $\frac{4}{3} \times \boxed{} = 1$ $\frac{3}{4}$

22. The number 37 has how many different factors?   2

23. $\dfrac{5}{5} \times \left(\dfrac{4}{4} - \dfrac{3}{3} \right) = 0$

24. To what decimal number is the arrow pointing?   5.4

25. How many edges does a cube have?   12

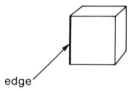

edge

LESSON
60

Subtracting Mixed Numbers with Borrowing

Subtracting mixed numbers is like subtracting fractions. To solve the problem

$$3\dfrac{2}{3} - 1\dfrac{1}{2}$$

we write the fractions with common denominators. Then we subtract the whole numbers and fractions and simplify when possible.

$$3\tfrac{2}{3} \times \tfrac{2}{2} = 3\tfrac{4}{6}$$
$$-1\tfrac{1}{2} \times \tfrac{3}{3} = 1\tfrac{3}{6}$$
$$\overline{\phantom{-1\tfrac{1}{2} \times \tfrac{3}{3} = }2\tfrac{1}{6}}$$

Sometimes when subtracting it is necessary to regroup. Wait until the fractions have been written with common denominators before regrouping.

example 60.1 $5\dfrac{1}{2} - 1\dfrac{2}{3} =$

solution   We write fractions with common denominators. We regroup if necessary. We simplify if possible.

$$5\tfrac{3}{6} = 4\tfrac{9}{6}$$
$$-1\tfrac{4}{6} = -1\tfrac{4}{6}$$
$$\overline{\phantom{-1\tfrac{4}{6} = -}3\tfrac{5}{6}}$$

practice
(See Practice
Set W in the
Appendix.)

a. 3
$-1\frac{2}{3}$
$\overline{1\frac{1}{3}}$

b. 5
$-3\frac{2}{5}$
$\overline{1\frac{3}{5}}$

c. $4\frac{1}{3}$
$-1\frac{2}{3}$
$\overline{2\frac{2}{3}}$

d. $5\frac{2}{5}$
$-3\frac{4}{5}$
$\overline{1\frac{3}{5}}$

e. $6\frac{1}{4}$
$-2\frac{1}{2}$
$\overline{3\frac{3}{4}}$

f. $4\frac{1}{3}$
$-1\frac{1}{2}$
$\overline{2\frac{5}{6}}$

g. $5\frac{1}{4}$
$-3\frac{1}{3}$
$\overline{1\frac{11}{12}}$

h. $6\frac{1}{2}$
$-1\frac{3}{4}$
$\overline{4\frac{3}{4}}$

i. $6\frac{1}{2}$
$-3\frac{5}{6}$
$\overline{2\frac{2}{3}}$

j. $8\frac{1}{6}$
$-4\frac{1}{3}$
$\overline{3\frac{5}{6}}$

k. $7\frac{2}{3}$
$-3\frac{5}{6}$
$\overline{3\frac{5}{6}}$

l. $8\frac{2}{3}$
$-5\frac{3}{4}$
$\overline{2\frac{11}{12}}$

problem set 60

Draw the fractions chart.

1. Convert the improper fraction $\frac{20}{6}$ to a mixed number with the fraction reduced. $3\frac{1}{3}$

2. A fathom is 6 feet. How many feet deep is water which is $2\frac{1}{2}$ fathoms?   15 ft.

3. After 3 days and 7,425 guesses, the queen guessed Rumpelstiltskin's name. What was the average number of names she guessed each day?   2475

*4. $5\frac{1}{2} - 1\frac{2}{3} = 3\frac{5}{6}$

5. $5\frac{1}{3} - 2\frac{1}{2} = 2\frac{5}{6}$

6. $1\frac{1}{2} + 2\frac{1}{3} + 3\frac{1}{4} = 7\frac{1}{12}$

7. $3\frac{3}{4} + 3\frac{1}{3} = 7\frac{1}{12}$

8. Compare: $\frac{2}{3} \bigcirc \frac{3}{5}$ >

9. $\frac{5}{6} \times 42 = 35$

10. $\frac{3}{8} \times \frac{2}{3} = \frac{1}{4}$

11. $\frac{3}{8} \div \frac{2}{3} = \frac{9}{16}$

12. $(4 - 0.4) \div 4 = 0.9$

13. $4 - (0.4 \div 4) = 3.9$

14. Which digit in 49.63 has the same place value as the 7 in 8.7?   6

15. Estimate the sum of $642.23 and $861.17 to the nearest hundred dollars. $1500

16. Which is closest to 10? 0.9, 6.1, 13.2   13.2

17. What number is next in this sequence? 100, 10, 1, ____
0.1 or $\frac{1}{10}$

18. The perimeter of a square is 1 foot. How many square inches cover its area?   9 sq. in.

19. Ten seconds is what fraction of one minute? $\frac{1}{6}$

20. ☐ − 15 = 7   22   **21.** List the factors of 50.
1, 2, 5, 10, 25, 50

22. By what name for 1 must $\frac{2}{3}$ be multiplied to form a fraction with a 15 in the denominator? $\frac{5}{5}$

23. What time is 5 hours and 15 minutes after 9:50 A.M.?
3:05 P.M.

24. How long is the line segment?
$1\frac{1}{8}$ in.

inches   1   2

25. Which word names this shape?   sphere
cylinder   cone   sphere   cube

LESSON
61

Prime Numbers

We have seen that whole numbers have factors.
The number six has four factors: 1, 2, 3, and 6.
The number seven has two factors: 1, and 7.
Whole numbers which have **exactly two factors** are called **prime numbers**. The only factors of a prime number are the number itself and 1. The first four prime numbers are 2, 3, 5, and 7. The number 2 is the only even prime number—all other even numbers have 2 as a factor. The number 2 is also the smallest prime number. The number 1 is not a prime number because it does not have exactly two factors; it has only one factor, 1.

example 61.1   The first four prime numbers are 2, 3, 5, and 7. What are the next four prime numbers?

solution Even numbers after 2 are not prime so we will consider the next several odd numbers: 9, 11, 13, 15, 17, and 19. We see that 9 has three factors (1, 3 and 9) and 15 has four factors (1, 3, 5, 15). The prime numbers are **11**, **13**, **17**, **19**.

practice Which is a prime number?   Which is not prime?

a. 21, 23, 25   23   d. 41, 42, 43   42

b. 31, 32, 33   31   e. 31, 41, 51   51

c. 43, 44, 45   43   f. 23, 33, 43   33

problem set 61 Draw the fractions chart.

1. In 1978 the total population of the world was approximately four billion, one hundred nineteen million. Write that number.   4,119,000,000

2. In music there are whole notes and half notes and quarter notes and eighth notes. How many quarter notes equal a whole note?   4

3. Don is 5 feet $2\frac{1}{2}$ inches tall. How many inches tall is that? $62\frac{1}{2}$ in.

*4. The first four prime numbers are 2, 3, 5, and 7. What are the next four prime numbers?   11, 13, 17, 19

5. Which of these is a prime number? 27, 35, 41, 63, 72, 84   41

6. The prices for three pairs of skates were $36.25, $41.50, and $43.75. What was the average price for a pair of skates? $40.50

7. $(44 \times 46) - (45 \times 45) = -1$

8. $3\frac{1}{4} + 4\frac{3}{8} = 7\frac{5}{8}$

9. $\frac{1}{2} + \frac{3}{4} + \frac{5}{8} = 1\frac{7}{8}$

10. $\frac{5}{6} - \frac{1}{2} = \frac{1}{3}$

11. $5\frac{1}{3} - 1\frac{1}{2} = 3\frac{5}{6}$

12. $\frac{1}{2} \times \frac{4}{5} = \frac{2}{5}$

13. $\frac{2}{3} \div \frac{1}{2} = 1\frac{1}{3}$

14. $1 - (0.2 - 0.03) = 0.83$

15. $0.14 \times 0.16 =$ 0.0224

16. $0.456 \div 6 =$ 0.076

17. $1.5 \div 0.04 =$ 37.5

18. One centimeter equals 10 millimeters. How many millimeters does 2.5 centimeters equal?   25 mm

19. List all the common factors of 18 and 24.   1, 2, 3, 6

20. What fraction of the group is shaded? $\frac{2}{5}$

21. If the perimeter of a square is 40 mm, what is the area of the square?   100 sq. mm

22. How many years were there from 15 B.C. to 87 A.D.?   101 years

23. Compare: $\dfrac{3}{3} \times \dfrac{2}{2} \bigcirc \dfrac{3}{3} \div \dfrac{2}{2}$ =

Use this chart of favorite sports of 100 people to answer questions 24 and 25.

24. How many more people favored baseball than favored football?   18

25. What fraction of the people favored baseball? $\frac{2}{5}$

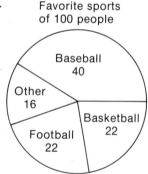

Favorite sports of 100 people

LESSON 62

Writing Whole Numbers and Mixed Numbers as Improper Fractions

Whole numbers like 2, 3, and 4 and mixed numbers like $1\frac{1}{2}$, $2\frac{2}{3}$, and $3\frac{3}{4}$ can be written as fractions with a numerator and denominator and no whole number part. Whole numbers and mixed numbers must be written as fractions to multiply or divide.

When a whole number or a mixed number is written as a fraction, it becomes an **improper fraction**. Improper fractions are equal to or greater than 1. You can identify an improper fraction by looking at it. The numerator is as large or larger than the denominator. Whole numbers can be written in fraction form simply by writing the whole number over a 1: $2 = \frac{2}{1}$, $3 = \frac{3}{1}$, $4 = \frac{4}{1}$.

Mixed numbers are made into fractions by dividing the whole number into the size pieces shown by the fraction:

$$1\frac{1}{2} = \frac{?}{2}$$

The picture shows $1\frac{1}{2}$ shaded circles. Dividing the whole circle into two halves, makes a total of three halves: $\frac{3}{2}$.

Below is a quick, mechanical method which can be learned to help us write mixed numbers as improper fractions. We multiply the denominator (bottom) by the whole number, then add the numerator (top). Follow the arrow in each example below.

$$1\,\overset{+1}{\underset{\times 2}{}} = \frac{3}{2} \qquad 2\,\overset{+2}{\underset{\times 3}{}} = \frac{8}{3} \qquad 3\,\overset{+3}{\underset{\times 4}{}} = \frac{15}{4}$$

practice Write these whole numbers and mixed numbers as improper fractions.

a. $3 \qquad \frac{3}{1}$

b. $3\frac{1}{2} \qquad \frac{7}{2}$

c. $4 \qquad \frac{4}{1}$

d. $2\frac{1}{3} \qquad \frac{7}{3}$

e. $1\frac{3}{4} \qquad \frac{7}{4}$

f. $6 \qquad \frac{6}{1}$

g. $5\frac{2}{3} \qquad \frac{17}{3}$

h. $2\frac{2}{3} \qquad \frac{8}{3}$

i. $6\frac{1}{4} \qquad \frac{25}{4}$

j. $3\frac{3}{10} \qquad \frac{33}{10}$

k. $12 \qquad \frac{12}{1}$

l. $12\frac{1}{2} \qquad \frac{25}{2}$

m. $2\frac{4}{5} \qquad \frac{14}{5}$

n. $1\frac{5}{6} \qquad \frac{11}{6}$

o. $11\frac{1}{4} \qquad \frac{45}{4}$

problem set 62 Draw the fractions chart.

1. What is the difference between the sum of 0.6 and 0.4 and the product of 0.6 and 0.4?   0.76

2. Mt. Whitney, the highest point in California, has an elevation of 14,494 feet above sea level. From there one can see Death Valley, the lowest point, which has an elevation of 282 feet **below** sea level. What is the difference in elevation between Mt. Whitney and Death Valley?   14,776 ft.

3. The dragon was 288 inches long. How many feet is this?
24 ft.

4. Write the whole number 10 as an improper fraction. $\frac{10}{1}$

5. Write the mixed number $4\frac{2}{3}$ as an improper fraction. $\frac{14}{3}$

6. What time is $2\frac{1}{2}$ hours after 10:15 A.M.?   12:45 P.M.

7. $(30 \times 15) \div (30 - 15) = $ 30   **8.** Compare: $\frac{5}{8} \bigcirc \frac{2}{3}$ <

9. $3\frac{2}{3} + 1\frac{1}{2} = 5\frac{1}{6}$ **10.** $\frac{6}{8} - \frac{3}{4} = $ 0   **11.** $6\frac{1}{4} - 5\frac{5}{8} = \frac{5}{8}$

12. $\frac{3}{4} \times \frac{2}{5} = \frac{3}{10}$ **13.** $\frac{3}{4} \div \frac{2}{5} = 1\frac{7}{8}$

14. $(1 - 0.4) \times (1 + 0.4) = $ 0.84   **15.** $0.4 \div 8 = $ 0.05

16. $8 \div 0.4 = $ 20   **17.** $\frac{3}{5}$ of $45 = $ 27

18. What number is next in this sequence? 0.2, 0.4, 0.6, 0.8, __1__

19. What is the tenth prime number?   29

20. What is the perimeter of the rectangle? $3\frac{3}{4}$ in.

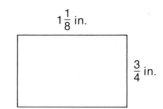

$1\frac{1}{8}$ in.

$\frac{3}{4}$ in.

21. How many square inches is 2 square feet?   288 sq. in.

22. Round $678.25 to the nearest ten dollars. $680

23. Write the name of this shape.   cone

24. A ton is 2000 pounds. How many pounds is $2\frac{1}{2}$ tons?
5000 lb.

25. Which arrow is pointing to 0.2 on the number line?   C

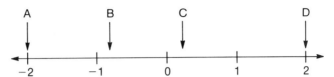

LESSON
63

Polygons

Polygons are shapes made up of straight sides. Polygons are named by the number of sides they have. The chart below names some common polygons.

SHAPE	NUMBER OF SIDES	NAME OF POLYGON
△	3	triangle
▭	4	quadrilateral
⬠	5	pentagon
⬡	6	hexagon
◯	8	octagon

Polygons in which all sides are equal and all angles are equal are called **regular polygons**.

example 63.1 What is the name of a polygon with four sides?

solution The answer is not "square." A square **does** have four sides, but not all four-sided polygons are squares. Check the chart until you are sure. The correct answer is **quadrilateral**. A square is a special type of quadrilateral.

example 63.2 A **regular octagon** has a perimeter of 96 inches. How long is each side?

solution An octagon has eight sides, and "regular" means all sides are the same length. Dividing, we find each side is **12 inches**. (Almost all of the red stop signs on our roads are regular octagons with sides 12 inches long.)

practice **a.** What is the name of a six-sided shape? hexagon
b. How many sides does a pentagon have? 5
c. Can a polygon have 19 sides? yes

problem Draw the fractions chart.
set 63

1. When the sum of 1.3 and 1.2 is divided by the difference of 1.3 and 1.2, what is the quotient? 25

2. William Shakespeare was born in 1564 and died in 1616. How many years did he live? 52 years

3. Robin Hood's arrow hit a target 45 yards away. How many feet did the arrow travel? 135 ft.

*4. What is the name of a polygon with four sides?
quadrilateral

*5. A regular octagon has a perimeter of 96 inches. How long is each side? 12 in.

6. $\$9 - 9¢ = \$8.91$ 7. $\dfrac{80 \times 80}{8 + 8} = 400$ 8. $5\dfrac{2}{3} + 3\dfrac{3}{4} = 9\dfrac{5}{12}$

9. $\dfrac{1}{2} + \dfrac{2}{3} + \dfrac{1}{4} = 1\dfrac{5}{12}$ 10. $\dfrac{9}{10} - \dfrac{1}{2} = \dfrac{2}{5}$ 11. $6\dfrac{1}{2} - 2\dfrac{7}{8} = 3\dfrac{5}{8}$

12. Compare: $2 \times 0.4 \bigcirc 2 + 0.4$ <

13. $4.8 \times 0.35 = 1.68$ 14. $4.8 \div 0.12 = 40$ 15. $1 \div 0.4 = 2.5$

16. Round the product of 0.33 and 0.38 to the nearest hundredth. 0.13

17. Multiply the length by the width to find the area of this rectangle.
$\frac{3}{8}$ sq. in.

$\frac{1}{2}$ in.

$\frac{3}{4}$ in.

18. Which is the twelfth prime number?   37

19. Write $3\frac{4}{5}$ as an improper fraction. $\frac{19}{5}$

20. ☐ $+ 0.2 = 1$ 0.8

21. Five minutes is what fraction of an hour? $\frac{1}{12}$

22. Name this shape.
cube

23. There are 100 centimeters to a meter. How many centimeters equal 2.5 meters?   250 cm

24. How long is the line segment? $3\frac{3}{8}$ in.

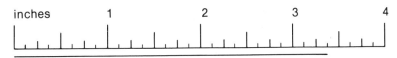

25. The ceiling in a house is about how many feet above the floor? (You may have to measure or ask.)   about 8 ft.

LESSON
64

Composite Numbers and Prime Factorization

All whole numbers greater than 1 are either prime numbers or **composite** numbers. A composite number has **more than two** factors. The numbers 2, 3, 5, 7 are prime. The numbers 4, 6, 8, 9 are composite. All composite numbers can be made by multiplying prime numbers together.

$$4 = 2 \times 2$$
$$6 = 2 \times 3$$
$$8 = 2 \times 2 \times 2 \text{ (not } 2 \times 4, \text{ because 4 is not prime)}$$
$$9 = 3 \times 3$$

When we write a composite number as a product of its prime factors, we have written the **prime factorization** of the number. **Factor trees** can help us write the prime factorization of a number.

example 64.1   Write the prime factorization of 36.

solution   Starting with 36, we break it into two factors. There are many choices (2 × 18, 3 × 12, 4 × 9, 6 × 6), but all will finally reduce to the factorization **2 × 2 × 3 × 3**. We show a factor tree for this.

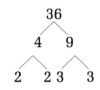

practice   Which are composite?   Draw factor trees for these numbers.
(See Practice   **a.** 19, 20, 23 20   Then write the prime factorizations.
Set X in the   **b.** 29, 31, 33 33   **e.** 10 2 × 5   **h.** 15 3 × 5
Appendix.)   **c.** 37, 39, 41 39   **f.** 12 2 × 2 × 3   **i.** 16 2 × 2 × 2 × 2
    d. 71, 72, 73 72   **g.** 14 2 × 7   **j.** 18 2 × 3 × 3

problem   Draw the fractions chart.
set 64

1. The total land area of the world is about fifty-seven million, two hundred eighty thousand square miles. Write that number. 57,280,000

2. The African white rhinoceros stood $6\frac{1}{2}$ feet high. How many inches is that? 78 in.

3. Jenny shot 10 free throws and made 6. What fraction of the shots did she make? $\frac{3}{5}$

*4. Draw a factor tree for 36. See drawing. 2 × 2 × 3 × 3

5. Which of these is a composite number? 21, 31, 41 21

6. What is the average of 3764, 2986, and 3429? 3393

7. $10,000 - (10,000 \div 16) = 9375$

8. $8\frac{1}{2} + 1\frac{1}{3} + 2\frac{1}{6} = 12$ 9. $\frac{1}{12} + \frac{1}{6} + \frac{1}{2} = \frac{3}{4}$

10. $15\frac{3}{4} - 2\frac{1}{8} = 13\frac{5}{8}$ 11. Compare: $\frac{1}{2} - \frac{1}{3} \bigcirc \frac{2}{3} - \frac{1}{2} =$

12. $\dfrac{3}{8} \times \dfrac{1}{3} = \dfrac{1}{8}$ **13.** $\dfrac{3}{8} \div \dfrac{1}{2} = \dfrac{3}{4}$

14. $1 - (0.2 + 0.48) = 0.32$ **15.** $\$0.08 \times 24 = \$1.92$

16. $0.0144 : 24 = 0.0006$ **17.** $20 \div 0.25 = 80$

18. What time is $2\frac{1}{2}$ hours before 1:15 P.M.?   10:45 A.M.

19. List **all** the **common** factors of 24 and 40.   1, 2, 4, 8

20. What is the name of a six-sided polygon?   hexagon

21. Write $3\frac{4}{7}$ as an improper fraction. $\frac{25}{7}$

22. What number is 100 less than 47?   **23.** $\dfrac{5}{6} \times \bigcirc = 1$ $\frac{6}{5}$
-53

24. How many **millimeters** long is the line segment?   50 mm

25. A meter is about one **big** step. About how many meters long is an automobile?   range about 5 or 6

LESSON
65

Multiplying Mixed Numbers

Mixed numbers are numbers with a whole number and a fraction written together, like $1\frac{2}{3}$. Mixed numbers can be multiplied and divided just like fractions. The fractions chart tells us the three steps to follow when multiplying mixed numbers.

1. Write the mixed numbers in fraction form.
2. Multiply numerators and denominators.
3. Simplify the answer by reducing and converting when possible.

example 65.1   Multiply: $1\dfrac{1}{2} \times 2\dfrac{2}{3}$

solution Write the mixed numbers as fractions. $\dfrac{3}{2} \times \dfrac{8}{3}$

Multiply numerators and denominators. $\dfrac{3}{2} \times \dfrac{8}{3} = \dfrac{24}{6}$

Simplify the answer. $\dfrac{24}{6} = \mathbf{4}$

example 65.2 Multiply: $2\dfrac{2}{3} \times 2$

solution Write in fraction form. $\dfrac{8}{3} \times \dfrac{2}{1}$

Multiply numerators and denominators. $\dfrac{8}{3} \times \dfrac{2}{1} = \dfrac{16}{3}$

Simplify the answer. $\dfrac{16}{3} = \mathbf{5\dfrac{1}{3}}$

practice
(See Practice Set Y in the Appendix.)

a. $1\dfrac{1}{2} \times \dfrac{2}{3} = 1$

b. $1\dfrac{2}{3} \times \dfrac{3}{4} = 1\dfrac{1}{4}$

c. $1\dfrac{1}{2} \times 1\dfrac{2}{3} = 2\dfrac{1}{2}$

d. $2\dfrac{1}{2} \times 2\dfrac{2}{3} = 6\dfrac{2}{3}$

e. $3\dfrac{1}{3} \times 1\dfrac{2}{3} = 5\dfrac{5}{9}$

f. $1\dfrac{2}{3} \times 3 = 5$

g. $3\dfrac{1}{2} \times 2 = 7$

h. $3 \times 1\dfrac{3}{4} = 5\dfrac{1}{4}$

i. $2\dfrac{2}{3} \times 4 = 10\dfrac{2}{3}$

j. $2\dfrac{3}{4} \times 2 = 5\dfrac{1}{2}$

problem set 65 Draw the fractions chart.

1. Write the mixed number $2\frac{2}{3}$ as an improper fraction. $\frac{8}{3}$

2. How many quarter notes equal a half note?   2

3. Railroad rails may weigh 155 pounds per yard and are often 33 feet long. How much would a 33-foot-long rail weigh?   1705 lb.

*4. $1\dfrac{1}{2} \times 2\dfrac{2}{3} = 4$

*5. $2\dfrac{2}{3} \times 2 = 5\dfrac{1}{3}$

6. Five numbers add up to 200. What is the average of the numbers?   40

7. $\dfrac{100 + 75}{100 - 75} = 7$ **8.** $1\dfrac{1}{5} + 3\dfrac{1}{2} = 4\dfrac{7}{10}$ **9.** $\dfrac{1}{3} + \dfrac{1}{6} + \dfrac{1}{12} = \dfrac{7}{12}$

10. $35\dfrac{1}{4} - 12\dfrac{1}{2} =$ **11.** $\dfrac{4}{5} \times \dfrac{1}{2} = \dfrac{2}{5}$ **12.** $\dfrac{4}{5} \div \dfrac{1}{2} = 1\dfrac{3}{5}$
$22\dfrac{3}{4}$

13. $3.75 - (1.2 \times 2) = 1.35$ **14.** $0.25 \div 4 = 0.0625$

15. $4 \div 0.25 = 16$

16. $\dfrac{1}{2} + \dfrac{1}{2} =$ (c)   (a) $\dfrac{1}{2} - \dfrac{1}{2}$ (b) $\dfrac{1}{2} \times \dfrac{1}{2}$ (c) $\dfrac{1}{2} \div \dfrac{1}{2}$

17. Write the prime factorization of 30. $2 \times 3 \times 5$

18. If five items cost 75¢, how much would seven items cost?
$1.05

19. Round $1.1675 to the nearest cent. $1.17

20. One side of a regular pentagon is 0.8 meter. What is the perimeter?   4 m

21. Twenty minutes is what fraction of an hour? $\frac{1}{3}$

22. The temperature dropped from 12° to −8°. This was a drop of how many degrees?   20°

23. What is the reciprocal of $2\frac{2}{3}$? $\frac{3}{8}$

Use the chart below to answer questions 24 and 25.

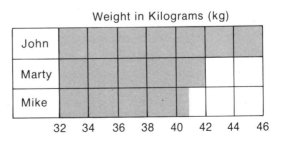

Weight in Kilograms (kg)

24. John weighs how much more than Mike?   5 kg

25. What is the average weight of the three boys?   43 kg

LESSON 66

Using Letters to Represent Unknowns

The sum is the answer to an addition problem. The numbers which are added together are called **addends**.

$$\text{addend} + \text{addend} = \text{sum}$$

The product is the answer to a multiplication problem. The numbers which are multiplied together are called **factors**.

$$\text{factor} \times \text{factor} = \text{product}$$

In earlier problem sets when an addend or factor was unknown, an empty box was in the problem, like

$$5 + \boxed{} = 8 \qquad \text{or} \qquad 3 \times \boxed{} = 12$$

Starting with this lesson, missing addends and factors may be shown by letters instead of boxes, like $5 + a = 8$ or $3n = 12$. Notice that when letters are used to show missing factors, the times sign ($\times$) is not used. The expression $3n$ means 3 times n. Any letter may stand for an unknown number.

example 66.1 $23 + f = 41 \qquad f =$

solution   Notice that we can find an unknown addend by subtracting the known addend from the sum ($41 - 23 = 18$). $f = \textbf{18}$

example 66.2 $8g = 96 \qquad g =$

solution   We can find an unknown factor by dividing the product by the known factor.

$$g = \frac{96}{8} \qquad \text{so} \qquad g = \textbf{12}$$

practice

$15 + a = 46$	$a = 31$	$5e = 60$	$e = 12$
$23 + b = 100$	$b = 77$	$6f = 114$	$f = 19$
$c + 14 = 50$	$c = 36$	$4g = 92$	$g = 23$
$d + 147 = 312$	$d = 165$	$3h = 123$	$h = 41$

problem set 66

Draw the fractions chart.

1. What is the sum of the first nine positive odd numbers?
81

2. A fathom is about 6 feet. A nautical mile is 1000 fathoms. A nautical mile is about how many feet?   6000 ft.

3. There are about 520 nine-inch-long noodles in a 1-pound package of spaghetti. Laid end to end, how many **feet** would the noodles in a package of spaghetti reach?   390 ft.

*4. $23 + f = 41$ $f = 18$ ***5.** $8g = 96$ $g = 12$

6. The combined length of four sticks is 172 inches. What is the average length of each stick?   43 in.

7. $10,000 - (4675 + 968) =$ 4357

8. $3\frac{1}{3} + 2\frac{3}{4} = 6\frac{1}{12}$

9. $\frac{7}{10} - \frac{1}{2} = \frac{1}{5}$

10. $4\frac{1}{4} - 2\frac{7}{8} = 1\frac{3}{8}$

11. $2\frac{2}{3} \times 3 = 8$

12. $1\frac{1}{3} \times 2\frac{1}{4} = 3$

13. $1\frac{1}{3} \div 2\frac{1}{4} = \frac{16}{27}$

14. $(2 \times 0.3) - (0.2 \times 0.3) =$ 0.54

15. $1.44 \div 60 = 0.024$

16. $6 \div 0.15 = 40$

17. Five dollars was divided evenly among four people. How much money did each receive? $1.25

18. What is the name of a **regular** quadrilateral?   square

19. Write the prime factorization of 40. $2 \times 2 \times 2 \times 5$

20. What is the area of the rectangle shown at right? $1\frac{1}{8}$ sq. in.

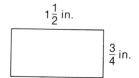

21. Thirty-six of the eighty-eight piano keys are black. What fraction of the piano keys are black? $\frac{9}{22}$

22. $(3 + 4) - (3 \times 4) = -5$

23. $1\frac{1}{2} \times \boxed{} = 1$ $\frac{2}{3}$

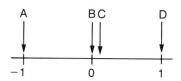

24. There are 1000 meters in a kilometer. How many meters are in 2.5 kilometers?   2500 m

25. Which arrow is pointing to 0.1 on the number line?   C

LESSON 67

Dividing Mixed Numbers

In another lesson we practiced multiplying mixed numbers. In this lesson we will practice dividing mixed numbers. Multiplying and dividing are on the same side of the fractions chart. The rules are very much alike. In fact, we have learned to change division problems into multiplication problems when working with fractions. The fractions chart tells us to follow these rules when dividing.

1. Write mixed numbers in fraction form.
2. Change the division problem into a multiplication problem by inverting the second fraction (multiply by the reciprocal of the divisor).
3. Multiply.
4. Simplify when possible.

example 67.1 $2\dfrac{2}{3} \div 1\dfrac{1}{2} =$

solution   Write the numbers as fractions. $\dfrac{8}{3} \div \dfrac{3}{2} =$

Invert the second fraction and change the symbol to multiplication. $\dfrac{8}{3} \times \dfrac{2}{3} =$

Multiply numerators and denominators. $\dfrac{8}{3} \times \dfrac{2}{3} = \dfrac{16}{9}$

Simplify when possible. $\dfrac{16}{9} = \mathbf{1\dfrac{7}{9}}$

practice
(See Practice Set Z in the Appendix.)

a. $\dfrac{3}{4} \div 1\dfrac{1}{2} = \dfrac{1}{2}$ **e.** $2\dfrac{1}{3} \div 1\dfrac{1}{2} = 1\dfrac{5}{9}$ **i.** $1\dfrac{1}{3} \div 2\dfrac{1}{2} = \dfrac{8}{15}$

b. $1\dfrac{1}{2} \div \dfrac{3}{4} = 2$ **f.** $4 \div 1\dfrac{1}{3} = 3$ **j.** $7 \div 1\dfrac{3}{4} = 4$

c. $1\dfrac{1}{2} \div 1\dfrac{1}{2} = 1$ **g.** $1\dfrac{1}{3} \div 4 = \dfrac{1}{3}$

d. $1\dfrac{1}{2} \div 2\dfrac{2}{3} = \dfrac{9}{16}$ **h.** $2\dfrac{1}{2} \div 1\dfrac{1}{3} = 1\dfrac{7}{8}$

problem set 67   Draw the fractions chart.

1. What is the difference between the sum of $\frac{1}{2}$ and $\frac{1}{4}$ and the product of $\frac{1}{2}$ and $\frac{1}{4}$? $\frac{5}{8}$

2. Bill ran a half mile in two minutes and fifty-five seconds. How many seconds is that?   175 sec.

3. The gauge of a railroad—the distance between the two tracks—is usually 4 feet $8\frac{1}{2}$ inches. How many inches is that? $56\frac{1}{2}$ in.

*4. $2\dfrac{2}{3} \div 1\dfrac{1}{2} = 1\dfrac{7}{9}$ 5. $1\dfrac{1}{2} \div 4 = \dfrac{3}{8}$

6. In six games Yvonne scored a total of 408 points. How many points per game did she average?   68

7. $\dfrac{1020}{24} = 42\dfrac{1}{2}$ 8. $5\dfrac{3}{8} + 1\dfrac{3}{16} = 6\dfrac{9}{16}$ 9. $3\dfrac{3}{5} + 2\dfrac{7}{10} = 6\dfrac{3}{10}$

10. $5\dfrac{1}{8} - 1\dfrac{1}{2} = 3\dfrac{5}{8}$ **11.** $3\dfrac{1}{3} \times 1\dfrac{1}{2} = 5$ **12.** $3\dfrac{1}{3} \div 1\dfrac{1}{2} = 2\dfrac{2}{9}$

13. What is the area of a rectangle which is 4 inches long and $1\frac{3}{4}$ inches wide?   7 sq. in.

14. $(3.2 + 1) - (0.6 \times 7) = 0$ **15.** $12.5 \div 0.4 = 31.25$

16. $0.375 \div 25 = 0.015$

17. $3.2 \times 10 =$ (b)
   (a) $32 \div 10$ (b) $320 \div 10$ (c) $0.32 \div 10$

18. Estimate the sum of 6416, 5734, and 4912 to the nearest thousand.   17,000

19. Write the prime factorization of 45. $3 \times 3 \times 5$

20. The perimeter of a square is 0.24 centimeter. How long is each side?   0.06 cm

21. What fraction of the months begin with the letter M? $\frac{1}{6}$

22. How many years was it from 500 B.C. to 500 A.D.?   999 years

23. By what name for 1 must $\frac{3}{4}$ be multiplied to form a fraction with a denominator of 100? $\frac{25}{25}$

24. List all the common factors of 8 and 12.   1, 2, 4

25. How long is the line segment? $1\frac{7}{8}$ in.

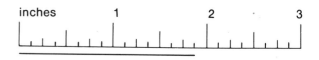

LESSON
68

Lines

The words **line**, **ray**, and **segment** have different meanings.

Line   Ray   Segment

A **line** continues in two directions without end.
A **ray** begins at one point and continues without end.
A **segment** is a part of a line.

    A **plane** in mathematics is a flat surface like a tabletop or a smooth sheet of paper. When two lines are drawn on the same plane, either they will cross at some point or they will not cross. When lines do not cross but stay the same distance apart, we say that the lines are **parallel**. When lines cross, we say they **intersect**. When they intersect and make square angles, or **right angles**, we call the lines **perpendicular**. A small square is often used to indicate perpendicular lines.

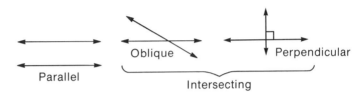

practice
  a. What do we call a part of a line? segment

  b. Is a beam of sunlight like a segment, a line, or a ray? ray

  c. What do we call lines in the same plane which do not intersect? parallel

Name the following.

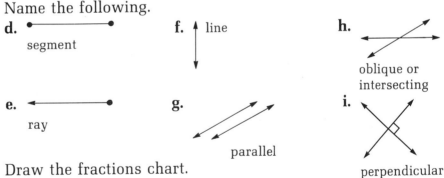

d. segment

e. ray

f. line

g. parallel

h. oblique or intersecting

i. perpendicular

problem set 68

Draw the fractions chart.

  1. What is the quotient if the dividend is $\frac{1}{2}$ and the divisor is $\frac{1}{8}$? 4

  2. The highest weather temperature recorded was 136° F in Africa. The lowest was −127° F in Antarctica. How many degrees difference is there between these temperatures? 263° F

  3. A dollar bill is about 6 inches long. Laid end to end, about how many **feet** would 1000 dollar bills reach? 500 ft.

4. What do we call a part of a line?   segment

5. What do we call a part of a line which begins at one point and continues without end in one direction?   ray

6. In 10 days Carla saved $27.50. On the average, how much had she saved each day? $2.75

7. $\dfrac{1 \times 2 \times 3 \times 4}{1 + 2 + 3 + 4} = 2\dfrac{2}{5}$

8. $\dfrac{2}{5} \times w = 1$ $w = \dfrac{5}{2}$

9. $3\dfrac{1}{2} + 2\dfrac{3}{4} + 1\dfrac{5}{8} = 7\dfrac{7}{8}$

10. $\dfrac{3}{4} - \dfrac{1}{3} = \dfrac{5}{12}$

11. $5\dfrac{3}{8} - 1\dfrac{3}{4} = 3\dfrac{5}{8}$

12. $2\dfrac{2}{3} \times 1\dfrac{1}{5} = 3\dfrac{1}{5}$

13. $1\dfrac{2}{3} \div 2 = \dfrac{5}{6}$

14. $\$10 - 9\cent = \$9.91$

15. What is the area of a square with sides 0.25 meter long?
0.0625 sq. m

16. $56.2 \div 100 = 0.562$

17. $20 \div 0.8 = 25$

18. Which digit in 97.68 has the same place value as the 2 in 132.45?   7

19. Write the prime factorization of 50.   2 × 5 × 5

20. What is the name of an eight-sided polygon?   octagon

21. There were 15 boys and 12 girls in the class. What fraction of the class was made up of girls? $\frac{4}{9}$

22. How many whole number spaces are between −3 and 8 on the number line?   11

23. What is the reciprocal of 2? $\frac{1}{2}$

24. There are 1000 grams in a kilogram. How many grams is 2.25 kilograms?   2250 g

25. About how many **millimeters** long is the line segment?
35 mm

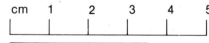

LESSON
69

Cancel

The only rule on the fractions chart left to practice is the **cancel** rule. **To cancel, we reduce before we multiply.** We cannot cancel when we add or subtract. We can only cancel when we multiply or when we have changed a division problem into a multiplication problem. To cancel, we may reduce any numerator with any denominator.

example 69.1 $\dfrac{5}{6} \times \dfrac{1}{5} =$

solution We could multiply then reduce, or reduce (cancel) then multiply. The 5 over 5 reduces (cancels) to 1 over 1.

$$\dfrac{\overset{1}{5}}{6} \times \dfrac{1}{\underset{1}{5}} = \dfrac{1}{6}$$

example 69.2 $\dfrac{5}{6} \div \dfrac{5}{2} =$

solution First we change division to multiplication by inverting $\frac{5}{2}$. Then we reduce 5 over 5 and 2 over 6. After canceling, we multiply.

$$\dfrac{\overset{1}{5}}{\underset{3}{6}} \times \dfrac{\overset{1}{2}}{\underset{1}{5}} = \dfrac{1}{3}$$

example 69.3 $\dfrac{9}{10} \times \dfrac{5}{6} =$

solution The larger the numbers are, the more helpful canceling becomes.

$$\dfrac{\overset{3}{9}}{\underset{2}{10}} \times \dfrac{\overset{1}{5}}{\underset{2}{6}} = \dfrac{3}{4}$$

practice
(See Practice
Sets)

Cancel before you multiply.   Invert, cancel, and multiply.

a. $\dfrac{3}{4} \times \dfrac{4}{5} = \dfrac{3}{5}$ **c.** $\dfrac{5}{6} \times \dfrac{3}{4} = \dfrac{5}{8}$ **e.** $\dfrac{2}{5} \div \dfrac{2}{3} = \dfrac{3}{5}$ **g.** $\dfrac{5}{6} \div \dfrac{5}{3} = \dfrac{1}{2}$

b. $\dfrac{2}{3} \times \dfrac{3}{4} = \dfrac{1}{2}$ **d.** $\dfrac{8}{9} \times \dfrac{9}{10} = \dfrac{4}{5}$ **f.** $\dfrac{2}{3} \div \dfrac{8}{9} = \dfrac{3}{4}$ **h.** $\dfrac{3}{5} \div \dfrac{9}{10} = \dfrac{2}{3}$

problem Draw the fractions chart.
set 69

1. Alaska was purchased from Russia in 1867 for seven million, two hundred thousand dollars. Write that amount.
$7,200,000

2. How many eighth notes equal a half note?   4

3. Children have 20 "baby teeth" which are later replaced by permanent teeth. What is the total cost to the Tooth Fairy for a child whose teeth have an average value of 75¢ each?
$15

*4. $\dfrac{5}{6} \times \dfrac{1}{5} = \dfrac{1}{6}$ *5. $\dfrac{5}{6} \div \dfrac{5}{2} = \dfrac{1}{3}$ *6. $\dfrac{9}{10} \times \dfrac{5}{6} = \dfrac{3}{4}$

7. What number is halfway between $\frac{1}{2}$ and 1 on the number line? $\frac{3}{4}$

8. $\dfrac{3}{4} + \dfrac{5}{6} = 1\dfrac{7}{12}$ 9. $3\dfrac{2}{3} + 4\dfrac{5}{6} = 8\dfrac{1}{2}$ 10. $7\dfrac{1}{8} - 2\dfrac{1}{2} = 4\dfrac{5}{8}$

11. $4.37 + 12.8 + 6 =$ 23.17

12. $0.46 \div 5 =$ 0.092   13. $60 \div 0.8 =$ 75

14. What is the average of 6.2, 4.3, 5, and 5.1?   5.15

15. $5 \times 6 =$ (c)
    (a) $50 \times 60 \div 10$ (b) $2 \div 60$ (c) $40 - (20 - 10)$

16. Round 16.1875 to the nearest hundredth.   16.19

17. $\dfrac{2}{3} + n = 1$ $n = \dfrac{1}{3}$ 18. $\dfrac{2}{3}m = 1$ $m = \dfrac{3}{2}$

19. Write the prime factorization of 60. $2 \times 2 \times 3 \times 5$

20. Segment AC is 47 mm. Segment AB is 19 mm. How long is segment BC?   28 mm

A   B   C

21. $37 - 102 =$ −65   22. $\dfrac{4}{5}$ of $60 =$ 48

23. There are 1000 milliliters in a liter. How many milliliters are in 3.8 liters?   3800 ml

Use the graph to answer questions 24 and 25.

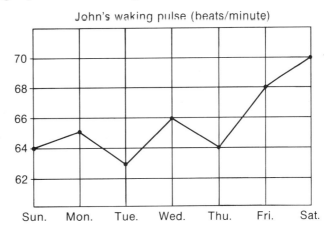

John's waking pulse (beats/minute)

24. When John woke on Saturday, his heart was beating how many more times per minute than it was on Tuesday?   7

25. On Monday, John took his pulse for 3 minutes before marking the graph. How many times did his heart beat in those 3 minutes?   195

LESSON 70

Multiplying Fractions— Three or More Factors

We have practiced using the rules for multiplying two fractions or two mixed numbers. We will use the same rules when we multiply three or more fractions or mixed numbers.

1. Write whole and mixed numbers in fraction form.
2. Cancel any factor of any numerator with any equal factor of any denominator.
3. Multiply all numerators and all denominators.
4. Simplify the answer if possible.

example 70.1 $\dfrac{2}{3} \times \dfrac{4}{5} \times \dfrac{3}{4} =$

solution   All amounts are already in fraction form. Cancel 4 over 4 and 3 over 3. Multiply numerators and denominators.

$$\frac{2}{\underset{1}{\cancel{3}}} \times \frac{\overset{1}{\cancel{4}}}{5} \times \frac{\overset{1}{\cancel{3}}}{\underset{1}{\cancel{4}}} = \frac{2}{5}$$

example 70.2 $1\frac{1}{4} \times 1\frac{1}{2} \times 2\frac{2}{3} =$

solution   Write in fraction form. Cancel where possible. Multiply and simplify.

$$\frac{5}{\underset{1}{\cancel{4}}} \times \frac{\overset{1}{\cancel{3}}}{\underset{1}{\cancel{2}}} \times \frac{\overset{2}{\cancel{8}}}{\underset{1}{\cancel{3}}} = \frac{5}{1} = 5$$

example 70.3 $\frac{3}{4} \times 1\frac{2}{3} \times 2 =$

solution   Write in fraction form. Cancel where possible. Multiply and simplify.

$$\frac{\overset{1}{\cancel{3}}}{\underset{2}{\cancel{4}}} \times \frac{5}{\underset{1}{\cancel{3}}} \times \frac{\overset{1}{\cancel{2}}}{1} = \frac{5}{2} = 2\frac{1}{2}$$

practice
(See Practice Sets Y and Z in the Appendix.)

a. $\frac{1}{2} \times \frac{2}{3} \times \frac{3}{4} = \frac{1}{4}$ d. $\frac{8}{9} \times \frac{6}{5} \times \frac{15}{16} = 1$ g. $1\frac{1}{6} \times 2 \times 4\frac{1}{2} = 10\frac{1}{2}$

b. $\frac{2}{3} \times \frac{1}{5} \times \frac{5}{6} = \frac{1}{9}$ e. $1\frac{1}{2} \times 2\frac{2}{3} \times \frac{3}{4} = 3$ h. $\frac{2}{3} \times \frac{3}{4} \times \frac{4}{5} \times \frac{5}{6} = \frac{1}{3}$

c. $\frac{5}{8} \times \frac{9}{10} \times \frac{4}{3} = \frac{3}{4}$ f. $1\frac{2}{3} \times 3\frac{1}{2} \times \frac{6}{7} = 5$

problem set 70   Draw the fractions chart.

1. What is the least common multiple of 6 and 10?   30

2. The highest point on land is Mt. Everest, which is 29,028 feet above sea level. The lowest point on land is the Dead Sea, which is 1299 feet below sea level. What is the difference in elevation between these two points?   30,327 ft.

3. The movie lasted 105 minutes. If it started at 1:15 P.M., at what time did it end?   3:00 P.M.

*4. $\frac{2}{3} \times \frac{4}{5} \times \frac{3}{4} = \frac{2}{5}$ *5. $1\frac{1}{4} \times 1\frac{1}{2} \times 2\frac{2}{3} = 5$

*6. $\frac{3}{4} \times 1\frac{2}{3} \times 2 = 2\frac{1}{2}$ 7. $4\frac{1}{2} \div 6 = \frac{3}{4}$

8. $6 + 3\frac{3}{4} + 2\frac{1}{2} =$ **9.** $5 - 3\frac{1}{8} = 1\frac{7}{8}$ **10.** $5\frac{1}{4} - 1\frac{7}{8} = 3\frac{3}{8}$

$12\frac{1}{4}$

11. $437 \times 86 =$ **12.** $\dfrac{5472}{18} = 304$ **13.** $15 \overline{)7505}$ 500 r5

37,582

14. $\$100 - \$10.87 =$ \$89.13   **15.** $(1 + 0.6) \div (1 - 0.6) =$ 4

16. $0.025 \overline{)40}$ 1600

17. What is the average of \$5.43, \$6, \$4.97, \$7.12?   \$5.88

18. $1.2f = 120$ $f =$ 100   **19.** $2 - \dfrac{2}{2} = 1$

20. Write the prime factorization of 64.
$2 \times 2 \times 2 \times 2 \times 2 \times 2$

21. The perimeter of a square is 6.4 meters. What is its area?
2.56 sq. m

22. Which is a ray?   (b)

(a)   (b)   (c)

23. What fraction is **not** shaded?

$\dfrac{3}{4}$

24. A centimeter is about this long ＿＿＿. About how many centimeters long is your little finger?   between 4 and 7 cm

25. What temperature is shown on the thermometer?   37°

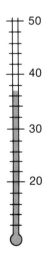

LESSON
71

Decimal Part of a Whole, Part 1

We have named part of a whole or part of a group by using common fractions. We may also name part of a whole or part of a group by using decimal fractions. Remember, the place value of the decimal number shows the number of parts into which the whole has been divided. We use one place after the decimal point (tenths) to name amounts divided into 10 parts. We use two places after the decimal point (hundredths) to name amounts divided into 100 parts, and so on.

example 71.1   What decimal part of the group is shaded?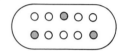

solution   Three of the ten parts are shaded. The common fraction $\frac{3}{10}$ is equal to the decimal fraction **0.3**.

example 71.2   What decimal part of a meter is 1 centimeter?

solution   A meter equals 100 centimeters. One centimeter is one-hundredth of a meter. Two places after the decimal point are used to name decimal amounts of wholes divided into 100 parts. We write **0.01**.

practice
(See Practice
Set AA in the
Appendix.)

a. Name the decimal amount shaded.   0.4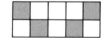
b. Name the decimal amount **not** shaded.   0.6
c. What decimal part of a dollar is a quarter?   0.25
d. What decimal part of a dime is a penny?   0.1
e. What decimal part of a century is a year?   0.01
f. What decimal part of a meter is 12 centimeters?   0.12
g. What decimal part of a decade is 3 years?   0.3
h. What decimal part of a dollar is a nickel?   0.05

**problem
set 71**

1. What is the average of 4.2, 2.61, and 3.6?   3.47

2. Four tablespoons equal $\frac{1}{4}$ of a cup. How many tablespoons would equal a full cup?   16 tablespoons

3. The temperature on the moon ranges from a high of 134° C to a low of about −170° C. This is a difference of how many degrees?   304° C

*4. What decimal part of the group is shaded?   0.3

*5. What decimal part of a meter is a centimeter?   0.01

6. What decimal part of a dollar is a nickel?   0.05

7. $\dfrac{1}{2} \times \dfrac{5}{6} \times \dfrac{4}{5} = \dfrac{1}{3}$

8. $3 \times 1\dfrac{1}{2} \times 2\dfrac{2}{3} = 12$

9. $\dfrac{3}{4} \div 2 = \dfrac{3}{8}$

10. $1\dfrac{1}{2} \div 1\dfrac{2}{3} = \dfrac{9}{10}$

11. $\dfrac{3}{5} + \dfrac{1}{2} = 1\dfrac{1}{10}$ **12.** $1 - \dfrac{7}{12} = \dfrac{5}{12}$ **13.** $3\dfrac{1}{3} - 2\dfrac{1}{2} = \dfrac{5}{6}$

14. $(1 + 2.3) - 0.45 = 2.85$ **15.** $0.12 \times 0.24 = 0.0288$

16. $0.6 \div 0.25 = 2.4$

17. Write the standard numeral for $(6 \times 10) + (4 \times \frac{1}{10}) + (3 \times \frac{1}{100})$.   60.43

18. Which is closest to 1?   (a) −1   (b) 0.1   (c) 10   (b)

19. What is the largest prime number which is less than 100?   97

20. $9 \times 4 = 6 \times \square$ 6

21. $a + 47 = 300$ $a = 253$

22. A square has a perimeter of 2 feet. How many inches long is each side?   6 in.

23. How many square inches of area does the square in question 22 occupy?   36 sq. in.

24. Seven tenths of the lights were on. What fraction of the lights were off? $\frac{3}{10}$

25. What is the total price for two Papa burgers, one Baby burger, one Big fries, and three drinks? $6.20

Papa burger	$1.45	Big fries	$0.75
Mama burger	1.05	Small fries	0.55
Baby burger	0.75	Drinks	0.60

LESSON 72

Writing Decimal Numbers as Fractions in Lowest Form

We can write parts of wholes or parts of groups two different ways. We can write them as common fractions or as decimal fractions. It is possible to change common fractions to decimal fractions and decimal fractions to common fractions. In this lesson we will review changing decimal fractions to common fractions.

A decimal fraction is a fraction with its denominator (10, 100, 1000, . . .) indicated by its place value. The decimal 0.3 means "three tenths" or $\frac{3}{10}$. We see that naming the decimal number names a common fraction as well. The decimal 0.2 is read "two tenths," which is the common fraction $\frac{2}{10}$. This time the fraction can be reduced. A fraction which can be reduced **should** be reduced. The decimal 0.2 equals the fraction $\frac{1}{5}$.

example 72.1   Write 0.5 as a common fraction.

solution   Read 0.5 as "five tenths." Write the fraction: $\frac{5}{10}$. Reduce the fraction when possible:

$$\frac{5}{10} = \frac{1}{2}$$

practice   Write the common fraction for these decimal fractions.

a. 0.3 $\frac{3}{10}$ **d.** 0.8 $\frac{4}{5}$ **g.** 0.25 $\frac{1}{4}$ **j.** 0.125 $\frac{1}{8}$

b. 0.4 $\frac{2}{5}$ **e.** 0.10 $\frac{1}{10}$ **h.** 0.75 $\frac{3}{4}$ **k.** 0.375 $\frac{3}{8}$

c. 0.6 $\frac{3}{5}$ **f.** 0.12 $\frac{3}{25}$ **i.** 0.24 $\frac{6}{25}$ **l.** 0.025 $\frac{1}{40}$

problem set 72

1. New York City uses about one billion, three hundred million gallons of water each day. Write that number.
1,300,000,000

2. Mark's temperature was 102° F. Normal body temperature is 98.6° F. How many degrees above normal was Mark's temperature? 3.4° F

3. Jill is reading page 42 of a 180-page book. If she must finish in three days, how many pages should she read each day?
46

*4. Write 0.5 as a common fraction. $\frac{1}{2}$

5. Write 0.35 as a common fraction. $\frac{7}{20}$

6. What decimal part of the shape is shaded? 0.4

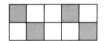

7. $\frac{3}{4} \times 2 \times 1\frac{1}{3} =$ 2

8. $(1000 - 625) \div 25 =$ 15

9. $3 + 2\frac{1}{3} + 1\frac{3}{4} =$ $7\frac{1}{12}$

10. $5\frac{1}{6} - 3\frac{1}{2} =$ $1\frac{2}{3}$

11. $\frac{3}{4} \div 1\frac{1}{2} =$ $\frac{1}{2}$

12. $\$1.43 - 68¢ =$ $\$0.75$

13. Compare: 0.3 ◯ 0.125 >

14. $6.3 \times 0.48 =$ 3.024

15. $0.175 \div 25 =$ 0.007

16. $7 \div 0.4 =$ 17.5

17. Which digit is in the ten-thousands' place in 123,456.78?
2

18. Arrange in order of size from least to greatest: $\frac{1}{2}, \frac{1}{10}, \frac{1}{4}$
$\frac{1}{10}, \frac{1}{4}, \frac{1}{2}$

19. Write the prime factorization of 72. $2 \times 2 \times 2 \times 3 \times 3$

20. $475 - y = 324$ $y =$ 151

21. AB is 16 cm. AC is 40 cm. How long is BC? 24 cm

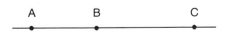

22. One-half of the area of the square is shaded. What is the area of the shaded region?   18 sq. in.

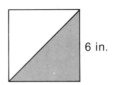

6 in.

23. $\dfrac{475 + 478 + 487}{3} = 480$ **24.** $47 - 200 = -153$

25. How long is the line segment?
$1\frac{3}{4}$ in.

inches   1   2   3

LESSON 73

Writing Common Fractions as Decimals

We can name fractional parts by using either common fractions or decimal fractions. We have practiced writing decimal fractions as common fractions. In this lesson we will practice writing common fractions as decimal fractions.

We learned in an earlier lesson that the fraction bar indicates division. The fraction $\frac{1}{2}$ means $2\overline{)1}$. Of course there is no whole number answer, which is the reason the answer is often left as a fraction. However, by attaching a decimal point and zeros we may do the division and write the answer as a decimal fraction.

$$\frac{1}{2} = 2\overline{)1} \qquad \begin{array}{r} 0.5 \\ 2\overline{)1.0} \\ \underline{1\,0} \\ 0 \end{array} \qquad \frac{1}{2} = 0.5$$

example 73.1   Convert the fraction $\dfrac{1}{4}$ to its decimal form.

solution The fraction $\frac{1}{4}$ also means 1 divided by 4, which is $4\overline{)1}$. By attaching a decimal point and zeros we may complete the division. We find that the fraction $\frac{1}{4}$ is equal to the decimal number 0.25.

$$\begin{array}{r} 0.25 \\ 4\overline{)1.00} \\ 8 \\ \hline 20 \\ 20 \\ \hline 0 \end{array}$$

practice
(See Practice Set BB in the Appendix.)

Convert each common fraction to its decimal form.

a. $\frac{1}{5}$ 0.2   d. $\frac{3}{4}$ 0.75   g. $\frac{1}{8}$ 0.125   j. $\frac{7}{20}$ 0.35

b. $\frac{2}{5}$ 0.4   e. $\frac{1}{10}$ 0.1   h. $\frac{3}{8}$ 0.375   k. $\frac{1}{25}$ 0.04

c. $\frac{3}{5}$ 0.6   f. $\frac{3}{10}$ 0.3   i. $\frac{1}{20}$ 0.05   l. $\frac{7}{25}$ 0.28

problem set 73

1. What is the sum of twenty-four thousand, eighty-six and one hundred nine thousand, two hundred twenty? 133,306

2. On a certain map, 1 inch represents a distance of 10 miles. How many miles apart are two towns which on the map are $2\frac{1}{2}$ inches apart? 25 mi.

3. Steve hit the baseball 400 feet. Tom hit the golf ball 300 yards. How many feet farther than the baseball did the golf ball travel? 500 ft.

*4. Convert the fraction $\frac{1}{4}$ to its decimal form. 0.25

5. To what decimal number is $\frac{4}{5}$ equal? 0.8

6. Write 0.24 as a reduced fraction. $\frac{6}{25}$

7. What decimal part of a dollar is 7 nickels? 0.35

8. $(100 - 20) \times (100 \div 20) =$ 400

9. $\frac{1}{2} + \frac{2}{3} + \frac{1}{6} =$ $1\frac{1}{3}$ 10. $3\frac{1}{4} - 1\frac{7}{8} =$ $1\frac{3}{8}$

11. $\frac{5}{8} \times \frac{3}{5} \times \frac{4}{5} =$ $\frac{3}{10}$ 12. $3\frac{1}{3} \times 3 =$ 10

13. $\dfrac{3}{4} \div \dfrac{1}{2} = 1\dfrac{1}{2}$ **14.** $(4 + 3.2) - 0.01 =$ 7.19

15. $\$4.50 \times 11 =$ $49.50   **16.** $\$0.96 \div 12 =$ $0.08

17. Estimate the product of 81 and 38.   3200

18. What is the **eighth** number in this sequence? 2, 4, 8, 16, 32, . . .   256

19. What is the least common multiple of 6, 8, and 12?   24

20. $24 + c + 96 = 150$ $c =$ 30

21. The perimeter of the rectangle is 48 inches. The width is 8 inches. What is the length?   16 in.

22. What is $\dfrac{1}{2}$ of 360?   180

23. Twenty-four of the three dozen bikers rode 10-speeds. What fraction of the bikers rode 10-speeds? $\frac{2}{3}$

24. How many years was it from 350 B.C. to 150 A.D.?
499 years

25. Which arrow is pointing to $\dfrac{3}{4}$?   B

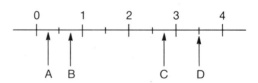

LESSON
74

Decimal Part of a Whole, Part 2

We may use a common fraction or a decimal fraction to name a fractional part of a whole or group. If the whole or the group is not divided into 10 parts or into 100 parts, the decimal part can still be named by converting the fraction into its decimal form.

example 74.1   What decimal part is shaded?

solution   Although the group does not have 10 parts, we can still name the shaded part with a decimal number. The fractional part shaded is $\frac{2}{5}$. We can convert the fraction $\frac{2}{5}$ into a decimal number by dividing:

$$5 \overline{)2.0} \quad \substack{0.4}$$

The decimal part shaded is **0.4**.

example 74.2   What decimal part is shaded?

solution   If the whole or group is not divided into 10 or 100 parts, then follow these steps.

1. Name the part with a common fraction.
2. Convert the common fraction to a decimal number.

In this case the fraction is $\frac{1}{4}$, which converts to the decimal **0.25**.

practice   **a.** What decimal part is **not** shaded?
0.6

b. What decimal part is **not** shaded?
0.75

c. What decimal part of a nickel is a penny?   0.2
d. What decimal part of a foot is 6 inches?   0.5

e. What decimal part is shaded?      0.375
f. What decimal part is **not** shaded?   0.625

problem   **1.** What is the reciprocal of three fifths? $\frac{5}{3}$
set 74

2. What time is one hour and thirty-five minutes after 2:30 P.M.?   4:05 P.M.

3. A 2-pound box of candy cost $8.00. What was the cost per ounce? (1 pound = 16 ounces.) $0.25

*4. What decimal part is shaded? 0.4

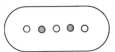

*5. What decimal part is shaded? 0.25

6. What decimal part is shaded? 0.5

7. Write $\dfrac{1}{8}$ as a decimal. 0.125

8. Write 0.8 as a reduced fraction. $\frac{4}{5}$

9. Write 0.12 as a reduced fraction. $\frac{3}{25}$

10. $\left(\dfrac{1}{2} + \dfrac{1}{3}\right) - \dfrac{1}{4} =$ $\frac{7}{12}$ **11.** $5 - 3\dfrac{1}{8} = 1\dfrac{7}{8}$ **12.** $3\dfrac{1}{2} \times 1\dfrac{1}{3} = 4\dfrac{2}{3}$

13. $3 - 1\dfrac{2}{3} = 1\dfrac{1}{3}$ **14.** $4 + $6.37 + 94¢ = $11.31

15. $1 - 0.95 =$ 0.05   **16.** $0.43 \times 2.6 =$ 1.118   **17.** $0.26 \div 5 =$ 0.052

18. Which digit in 4.87 has the same place value as the 9 in 0.195? 7

19. $0.65 \div 10 =$ 0.065   **20.** $12 \div 0.04 =$ 300

21. What is the greatest common factor of 18 and 30? 6

22. $25N = 1000.$ $N =$ 40   **23.** Compare: $0.3 \bigcirc 0.25$ >

24. The area of a square is 81 square feet. What is its perimeter? 36 ft.

25. How many teaspoons equal one gallon?   768 teaspoons

3 teaspoons	= 1 tablespoon
16 tablespoons	= 1 cup
2 cups	= 1 pint
2 pints	= 1 quart
4 quarts	= 1 gallon

LESSON
75

Comparing Fractions— Decimal Form

Fractions can be compared when their denominators are the same. The fractions $\frac{5}{8}$ and $\frac{4}{5}$ cannot be compared in their present form because their denominators are not the same. To compare these fractions they must be named with common denominators. One way to write these fractions with common denominators is to write them as decimal numbers with the same number of decimal places.

$$\frac{5}{8} = 8\,\overline{)\,5.000} = 0.625 \qquad \frac{4}{5} = 5\,\overline{)\,4.000} = 0.800$$

$$\frac{5}{8}\bigcirc\frac{4}{5} \text{ is the same as } 0.625\bigcirc0.800$$

example 75.1   Compare: $\dfrac{3}{5}\bigcirc\dfrac{5}{8}$ (Convert both numbers to decimal numbers.)

solution   Fractions can be compared when their denominators are equal. One way to write fractions with common denominators is to write them as decimal numbers with the same number of places. The fraction $\frac{3}{5}$ is 0.6 and the fraction $\frac{5}{8}$ is 0.625. Comparing $\frac{3}{5}$ and $\frac{5}{8}$ is like comparing 0.600 and 0.625, so the fraction $\frac{5}{8}$ is greater than $\frac{3}{5}$.

$$\frac{3}{5} < \frac{5}{8} \text{ because } 0.600 < 0.625$$

practice   Compare these fractions by first changing them to decimal numbers with the **same** number of decimal places.

a. $\frac{1}{4} \bigcirc \frac{3}{5}$ 0.25 < 0.60   **e.** $\frac{3}{20} \bigcirc \frac{1}{8}$ 0.150 > 0.125   **i.** $\frac{3}{50} \bigcirc \frac{1}{20}$ 0.06 > 0.05

b. $\frac{3}{4} \bigcirc \frac{7}{10}$ 0.75 > 0.70   **f.** $\frac{1}{5} \bigcirc \frac{1}{4}$ 0.20 < 0.25   **j.** $\frac{10}{25} \bigcirc \frac{2}{5}$ 0.4 = 0.4

c. $\frac{3}{8} \bigcirc \frac{2}{5}$ 0.375 < 0.400   **g.** $\frac{7}{10} \bigcirc \frac{2}{5}$ 0.7 > 0.4

d. $\frac{3}{5} \bigcirc \frac{1}{2}$ 0.6 > 0.5   **h.** $\frac{3}{8} \bigcirc \frac{9}{25}$ 0.375 > 0.360

problem   **1.** When the product of 8 and 6 is divided by the difference
set 75   of 8 and 6, what is the quotient?   24

2. How many eighth notes equal a quarter note?   2

3. It is said that one year of a dog's life is like 7 years of a human's life. In that case, a dog that is 13 years old is like a human that is how many years old?   91

*4. Compare: $\frac{3}{5} \bigcirc \frac{5}{8}$ (Convert both numbers to decimal numbers).   <

5. What decimal part is shaded?
0.75

6. Write $\frac{5}{8}$ as a decimal.   0.625

7. Write 0.45 as a reduced fraction. $\frac{9}{20}$

8. Write 0.04 as a reduced fraction. $\frac{1}{25}$

9. Divide 1000 by 9 and write the quotient with a fraction.
$111\frac{1}{9}$

10. $6\frac{1}{3} + 3\frac{1}{4} + 2\frac{1}{2} = 12\frac{1}{12}$ **11.** $\frac{5}{8} + \frac{5}{6} = 1\frac{11}{24}$

12. $2\frac{1}{2} \times 3\frac{1}{3} = 8\frac{1}{3}$ **13.** $5 \div 2\frac{1}{2} = 2$

14. $6.7 + 0.48 + 5 = 12.18$ **15.** $12 - 4.75 = 7.25$

16. $0.35 \times 0.45 = 0.1575$ **17.** $4.3 \div 100 = 0.043$

18. Arrange in order of size from least to greatest: 0.3, 0.25, 0.313   0.25, 0.3, 0.313

19. Estimate the sum of 3,926 and 5,184 to the nearest thousand.
9000

20. List all the prime numbers between 40 and 50.
41, 43, 47

21. $476 - w = 284$ $w = 192$

22. What is the perimeter of the triangle?
$2\frac{2}{5}$ in.

23. $365 - 500 = -135$

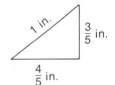

24. How many **millimeters** long is the line segment?
40 mm

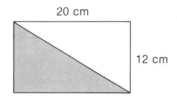

25. One-half of the area of the rectangle is shaded. What is the area of the shaded region?   120 sq. cm

20 cm

12 cm

LESSON
76

Fractional Part of a Number

The **denominator** of a fraction shows into how many parts the whole or group has been or should be divided. The **numerator** tells how many of those parts are to be counted. The difference between the numerator and denominator is how many parts are not counted.

The statement "$\frac{3}{5}$ of 20" means that 20 has been divided into five equal parts. Three of those parts are counted. Two of those parts $(5 - 3)$ are not counted.

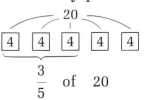

example 76.1   There are 28 students in the class. Three fourths of them are boys.

   1. Into how many parts was the class divided?
   2. How many are in each part?
   3. How many parts are boys?
   4. How many boys are in the class?
   5. How many parts are girls?
   6. How many girls are in the class?

solution   The class was divided into **four parts** with **seven** students in each part. **Three** parts are boys for a total of **twenty-one** boys. **One** part is girls for a total of **seven** girls.

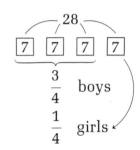

practice   Three-eighths of the 40 little engines could.
(See Practice   **a.** Into how many parts was the group divided?   8
Set CC in the   **b.** How many engines are in each part?   5
Appendix.)   **c.** How many parts could?   3
   d. How many engines could?   15
   e. How many parts could not?   5
   f. How many engines could not?   25

problem   1. What is the quotient when the decimal number ten and
set 76   five tenths is divided by four hundredths?   262.5

   2. The weight of an object on the moon is $\frac{1}{6}$ of its weight on earth. A person weighing 114 pounds on earth would weigh how much on the moon?   19 lb.

   3. Cupid shot 24 arrows and hit 6 targets. What fraction of his shots hit the target? $\frac{1}{4}$

*There are 28 students in the class. Three fourths of them are boys. Use this information to answer questions 4 to 9.

   4. Into how many parts was the class divided?   4

   5. How many are in each part?   7

6. How many parts are boys? 3

7. How many boys are in the class? 21

8. How many parts are girls? 1

9. How many girls are in the class? 7

10. What decimal part is shaded?
0.375

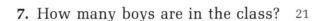

11. What fraction is equal to the decimal number 0.6? $\frac{3}{5}$

12. Compare: $0.35 \bigcirc \frac{7}{20}$ =

13. $\frac{1}{2} + \frac{2}{3} = 1\frac{1}{6}$ **14.** $3\frac{1}{5} - 1\frac{3}{5} = 1\frac{3}{5}$ **15.** $\frac{1}{2} + \frac{3}{4} + \frac{7}{8} = 2\frac{1}{8}$

16. $3 \times 1\frac{1}{3} = 4$ **17.** $3 \div 1\frac{1}{3} = 2\frac{1}{4}$ **18.** $1\frac{1}{3} \div 3 = \frac{4}{9}$

19. What is the perimeter of the rectangle?
4.8 cm

20. What is the area of the rectangle?
1.35 sq. cm

21. Which digit in 6734.2198 is in the ones' place? 4

22. $3.6 + a = 4.15$ $a = 0.55$

23. Round $357.64 to the nearest dollar. $358

24. How long is the line segment?
$1\frac{3}{8}$ in.

25. What time is one hour and fourteen minutes before noon?
10:46 A.M.

LESSON
77

Decimal Part of a Number

The **place value** of the last digit of a decimal fraction shows into how many parts a whole or group has been or should be divided. The **digits** show how many parts are counted. The difference between the total parts and counted parts is the uncounted parts.

The statement "0.4 of 20" means that 20 was divided into ten parts. Four of the parts are counted. Six parts are not counted.

20

2	2	2	2	2
2	2	2	2	2

0.4 of 20 = 8

example 77.1   If 0.8 of the 60 lights are "on" . . .

1. Into how many parts have the 60 lights been divided?
2. How many lights are in each part?
3. How many parts are "on"?
4. How many lights are "on"?
5. How many parts are "off"?
6. How many lights are "off"?

solution   The place value of the 8 shows that the lights were divided into **ten** groups, which makes **six** lights in each group. **Eight** of the groups were "on," which is **48** lights, and **two** of the groups were "off," which is **12** lights.

60

6	6
6	6
6	6
6	6
6	6

"on"

"off"

practice
(See Practice
Set CC in the
Appendix.)

If 0.6 of the 30 students were girls:
a. Into how many parts were the students divided?   10
b. How many students were in each part?   3
c. How many parts are girls?   6
d. What is the number of girls?   18
e. How many parts are boys?   4
f. What is the number of boys?   12

problem   **1.** What is the difference when the product of $\frac{1}{2}$ and $\frac{1}{2}$ is
set 77   subtracted from the sum of $\frac{1}{2}$ and $\frac{1}{2}$? $\frac{3}{4}$

2. The claws of a Siberian tiger are 10 centimeters long. How
many millimeters long is that?   100 mm

3. Sue was thinking of a number between 40 and 50 which
is a multiple of 3 and 4. Of what number was she thinking?
48

*If 0.8 of the lights are "on" . . .

4. Into how many parts have the 60 lights been divided?   10

5. How many lights are in each part?   6

6. How many parts are "on"?   8

7. How many lights are "on"?   48

8. How many parts are "off"?   2

9. How many lights are "off"?   12

10. What decimal part is shaded?
0.25

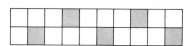

11. What fraction is equal to the decimal number 0.15? $\frac{3}{20}$

12. Compare: $\dfrac{3}{5}$ ◯ 0.35   >

13. $\dfrac{5}{6} - \dfrac{1}{2} = \dfrac{1}{3}$ **14.** $3\dfrac{1}{4} - 1\dfrac{3}{4} = 1\dfrac{1}{2}$ **15.** $\dfrac{1}{2} + \dfrac{2}{3} + \dfrac{5}{6} = 2$

16. $1\dfrac{1}{2} \times 2\dfrac{2}{3} = 4$ **17.** $1\dfrac{1}{2} \div 2\dfrac{2}{3} = \dfrac{9}{16}$ **18.** $2\dfrac{2}{3} \div 1\dfrac{1}{2} = 1\dfrac{7}{9}$

19. What is the perimeter of the square?
2 in.

20. What is the area of the square?
$\frac{1}{4}$ sq. in.

21. Name the place value of the 6 in 4397.6185.   tenths

22. $3a = 24$ $a = 8$

23. Round 1.3579 to the hundredths' place.   1.36

24. How many inches is $2\frac{1}{2}$ feet?   30 in.

25. Which arrow is pointing to 0.1?   C

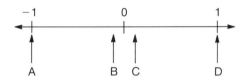

LESSON
78

Classifying Quadrilaterals

Quadrilaterals are polygons with four sides.
    Quadrilaterals are classified in the following way.

SHAPE	CHARACTERISTIC	NAME
	No sides parallel	trapezium
	One pair of parallel sides	trapezoid
	Two pairs of parallel sides	parallelogram
	Parallelogram with equal sides	rhombus
	Parallelogram with right angles	rectangle
	Rectangle with equal sides	square

Notice that a square is a special kind of rectangle which is a special kind of parallelogram which is a special kind of quadrilateral which is a special kind of polygon. A square is also a special kind of rhombus.

example 78.1   Is the following statement true or false?
"All parallelograms are rectangles."

solution   A rectangle **is** a special kind of parallelogram, so some parallelograms are rectangles, and all rectangles are parallelograms.
However, there are some parallelograms which are not rectangles, so the statement is **false**.

practice   State whether each statement is true or false.
    a. All quadrilaterals are four-sided shapes.   true
    b. Some parallelograms are trapezoids.   false
    c. Every square is a rhombus.   true
    d. Every rhombus is a square.   false
    e. Every square is a rectangle.   true
    f. Every rectangle is a square.   false

problem   **1.** What is the reciprocal of $2\frac{1}{2}$? $\frac{2}{5}$
set 78

    2. If 500 calories adds 1 pound to your weight, how many
1000-calorie banana splits would you need to eat to gain
10 pounds?   5

    3. Uncle Bill was 38 when he started his job. He worked for
33 years. How old was he when he retired?   71

    *4. Is the following statement true or false?
"All parallelograms are rectangles."   false

    5. "All four-sided shapes are quadrilaterals." (True or false?)
true

    6. If 0.9 of the 30 students were right-handed, then how many
students were left-handed?   3

    7. If $\frac{3}{4}$ of the 24 runners finished the race, how many runners
did not finish the race?   6

    8. $(1000 - 124) \div 12 =$ 73   **9.** $6.42 + 12.7 + 8 =$ 27.12

    10. $10 - 9.87 =$ 0.13   **11.** $1.2 \times 0.12 =$ 0.144

    12. $0.288 \div 24 =$ 0.012   **13.** $64 \div 0.08 =$ 800

14. $3\dfrac{1}{3} + 2\dfrac{3}{4} = 6\dfrac{1}{12}$ **15.** $\dfrac{5}{6} - \dfrac{1}{4} = \dfrac{7}{12}$

16. $3\dfrac{1}{3} \times \dfrac{1}{5} = \dfrac{2}{3}$ **17.** $2\dfrac{1}{2} \div 3 = \dfrac{5}{6}$

18. The perimeter of a square is 400 cm. What is its area?
10,000 sq. cm

19. Write the decimal numeral for $(9 \times 10) + (6 \times 1) + (3 \times \frac{1}{100})$.   96.03

20. What is the **tenth** number in this sequence? 6, 12, 18, 24, 30, . . .   60

21. Which is closest to zero? -2, 0.2, 1, $\dfrac{1}{2}$ 0.2

22. Estimate the product of 67 and 73.   4900

23. Round 26.7 to the nearest whole number.   27

24. What number is halfway between 37 and 73?   55

25. To what **decimal** number is the arrow pointing?   10.2

**LESSON
79**

Area of a Parallelogram

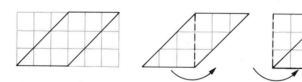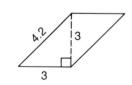

Above are four pictures of the same parallelogram. The area
of a parallelogram is the total number of squares enclosed by

its sides. How many squares are enclosed by the sides of this parallelogram? We can count 6 full squares and 6 half squares. Six half squares will make 3 whole squares, so by counting we find the area of the parallelogram is a total of 9 square units.

The second and third pictures show another way to find the area. Imagine that the parallelogram is cut along the dotted line. Move the part on the left to fill the space under the part on the right. We see that the area of the parallelogram is the same as the area of a 3 by 3 square, which is 9 square units.

Looking at the fourth picture, what numbers must be multiplied to give an answer of 9 square units? We see that it is **not** the base and a side which must be multiplied, but it is the **base** and the **height** which must be multiplied to find the area. The height is **perpendicular** to the base. **To find the area of a parallelogram, multiply the base and height.**

practice   Find the area of each parallelogram.

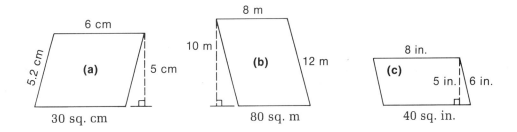

problem set 79

1. What is the average of 96, 49, 68, and 75?   72

2. The average depth of the ocean beyond the edges of the continents is $2\frac{1}{2}$ miles. How many feet is that? (1 mile = 5280 ft.)   13,200 ft.

3. The 168 girls who signed up for soccer were divided into 12 teams. How many players were on each team?   14

*4. What is the area of the parallelogram?
    9 sq. cm

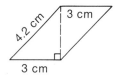

5. Name this shape.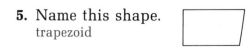
trapezoid

6. "All squares are rectangles." (True or false?)   true

7. If 0.8 of the class was present, then what decimal part of the class was absent?   0.2

8. If $\frac{1}{6}$ of the one-hour show was taken up with commercials, then how many minutes did the commercials last?
10 min.

9. Compare: $0.5 \bigcirc \dfrac{3}{4}$ $<$

10. Write 0.4 as a reduced common fraction. $\frac{2}{5}$

11. Write $\dfrac{1}{8}$ as a decimal fraction.   0.125

12. $6.3 \times 0.36 =$
2.268

13. $0.36 \div 5 =$
0.072

14. $63 \div 0.9 = 70$

15. $\dfrac{5}{8} + \dfrac{1}{2} = 1\dfrac{1}{8}$

16. $\dfrac{5}{8} - \dfrac{1}{4} = \dfrac{3}{8}$

17. $2\dfrac{1}{2} \times 1\dfrac{1}{3} = 3\dfrac{1}{3}$

18. $3 \div \dfrac{1}{2} = 6$

19. $1\dfrac{1}{2} \div 3 = \dfrac{1}{2}$

20. What is the area of the parallelogram?   156 sq. mm

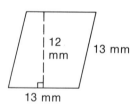
12 mm
13 mm
13 mm

21. Round 0.4287 to the hundredths' place.   0.43

22. How many centimeters is 15 meters?   1500 cm

23. $4 - a = 2.6$ $a = 1.4$

Use the graph of sugar in breakfast cereals to answer questions 24 and 25.

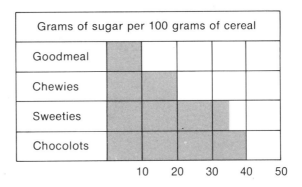

24. Sweeties contains about how many grams of sugar per 100 grams of cereal?   35 grams

25. Fifty grams of Chocolots would contain about how many grams of sugar?   20 grams

LESSON
80

Other Multiplication Forms

We are familiar with showing multiplication by the "times" sign, ×. In this lesson we will learn two other ways of showing multiplication.

One way of showing multiplication is by placing a dot between the numbers being multiplied. "Five times three" may be written $5 \cdot 3$. Notice that the multiplication dot is at the middle level, not down at the bottom of the numbers like a decimal point would be. "Five times n" could be written $5 \cdot n$.

Another way of showing multiplication is to write the amounts next to each other with no sign between. This is the most common way of showing multiplication in later math courses. "Five times n" can be written $5n$. To show "a times b" we can write ab. When we write the multiplication of two **numerals** in this way, we must use parentheses to avoid confusion. It would not be clear whether 53 meant "5 times 3" or "fifty-three" unless we wrote 5 times 3 as 5(3). Using parentheses allows us to write multiplied numerals together with no other sign. This multiplication may be shown as 5(3) or (5)3 or (5)(3). There is no difference in meaning.

practice
a. 6 · 4   24
b. 6(5)   30
c. (6)6   36
d. (6)(7)   42

e. 2 · 3 · 4   24
f. 2(3)5   30
g. 1 · 1 · 1 · 1   1
h. (2)(2)(2)(2)   16

problem set 80

1. What is the sum of one million, eighty-seven thousand, eight hundred twelve and nine hundred twenty thousand, six hundred five?   2,008,417

2. Jim was thinking of a prime number between 75 and 100 which did **not** have a 9 as one of its digits. Of what number was he thinking?   83

3. Carol cut $2\frac{1}{2}$ inches off her hair three times last year. How much longer would her hair have been if she had not cut it? $7\frac{1}{2}$ in.

4. 5 · 3 = 15

5. 12(24) = 288

6. If $n = 8$ then what is the value of $5n$?   40

7. What is the area of the parallelogram?
   600 sq. m

8. What is the perimeter of the parallelogram?   100 m

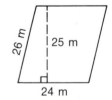

9. "Some rectangles are trapezoids." (True or false?)   false

10. What decimal part is **not** shaded?
    0.4

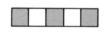

11. Which of these is greatest? $\frac{1}{2}, \frac{1}{5}, 0.4$ $\frac{1}{2}$

12. What decimal number is equal to $\frac{4}{25}$?   0.16

13. The decimal number 0.45 is equal to what reduced fraction?
    $\frac{9}{20}$

14. $(0.4 + 3) \div 2 =$ 1.7

15. $(10 - 0.1) \times 0.1 =$ 0.99

16. $\frac{5}{8} + \frac{3}{4} = 1\frac{3}{8}$

17. $3 - 1\frac{1}{8} = 1\frac{7}{8}$

18. $4\frac{1}{2} - 1\frac{3}{4} = 2\frac{3}{4}$

19. $\frac{5}{6} \times \frac{3}{5} = \frac{1}{2}$

20. $4\frac{1}{2} \times 1\frac{1}{3} = 6$

21. $3\frac{1}{3} \div 1\frac{2}{3} = 2$

22. The perimeter of a square is 2 meters. How many centimeters long is each side? (1 meter = 100 centimeters.)
50 cm

23. $w - 72 = 36$ $\qquad w = 108$

24. What time is two and one-half hours after 10:40 A.M.
1:10 P.M.

25. How long is the line segment?
$2\frac{5}{8}$ in.

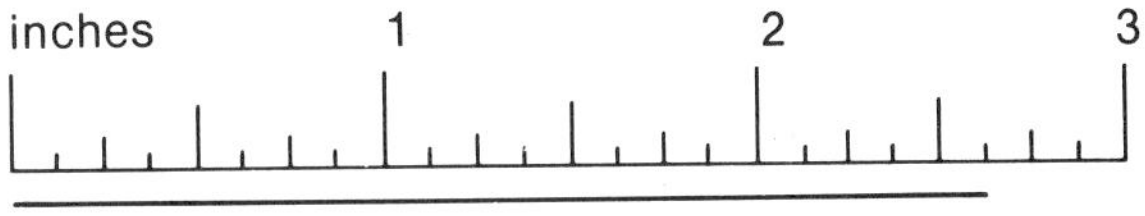

LESSON 81 Percent, Part 1

There are three ways to name fractions:

1. as common fractions,
2. as decimals, and
3. as percents.

We have practiced writing numbers as fractions and as decimals. In this lesson we will write numbers as **percents**.

A percent is actually a fraction with a denominator of 100. Instead of writing the denominator 100, we write a percent sign (%). A number with a percent sign is exactly equal to that number with a denominator of 100. **The number 25% means $\frac{25}{100}$.**

example 81.1 Write $\frac{3}{100}$ as a percent.

solution A percent is a fraction with a denominator of 100. Instead of writing the denominator, a percent sign is used. We write $\frac{3}{100}$ as **3%**.

example 81.2   Write 99% as a fraction.

solution   A percent is a fraction with a denominator of 100. The percent sign can be written as a denominator of 100. We write 99% as the fraction $\frac{99}{100}$.

example 81.3   Of the 100 students who took the test, 23 earned A's. What percent of the students earned A's?

solution   As a fraction we would write $\frac{23}{100}$. The percent which equals $\frac{23}{100}$ is **23%**.

practice   Write as fractions.   Write as percents.

   a. 27% $\frac{27}{100}$

   b. 91% $\frac{91}{100}$ **d.** $\frac{31}{100}$ **e.** $\frac{1}{100}$ **f.** $\frac{57}{100}$

   c. 7% $\frac{7}{100}$ 31%   1%   57%

problem set 81

1. Write the decimal numeral twenty-one and five hundredths.   21.05

2. Tennis balls are sold in cans containing 3 balls. What would be the total cost of buying one dozen tennis balls if the price per can was $2.49? $9.96

3. A cubit is 18 inches. If David was 4 cubits tall, how many feet tall was he?   6 ft.

*4. Write $\frac{3}{100}$ as a percent.   3%

*5. Write 99% as a fraction. $\frac{99}{100}$

*6. Of the 100 students who took the test, 23 earned A's. What percent of the students earned A's?   23%

7. Write 17% as a fraction. $\frac{17}{100}$

8. Write $\frac{23}{100}$ as a percent.   23%

9. $3\frac{5}{6} + 2\frac{1}{3} = 6\frac{1}{6}$ 10. $3\frac{1}{4} - 1\frac{5}{8} = 1\frac{5}{8}$ 11. $\frac{1}{2} + \frac{1}{5} + \frac{1}{10} = \frac{4}{5}$

12. $1\dfrac{4}{5} \times 1\dfrac{2}{3} = 3$ **13.** $6 \div 1\dfrac{1}{2} = 4$

14. What decimal part of the group is shaded?   0.625

15. Write 0.45 as a reduced common fraction. $\frac{9}{20}$

16. $(1 - 0.84) \times 0.14 = 0.0224$ **17.** $0.15 \div 12 = 0.0125$

18. $6 \div 0.12 = 50$

19. Which digit in 6.3457 has the same place value as the 8 in 128.90?   6

20. Estimate the product of 20, 39, and 41.   32,000

21. $5n = 45$ $n = 9$

22. What is the area of the parallelogram?
500 sq. mm

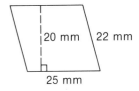

23. $\dfrac{(42)(12)}{14}$ 36

24. $514 - 786 = -272$

25. How long is the line segment?
$1\frac{5}{8}$ in.

LESSON
82

Percent, Part 2

Parts of wholes and parts of groups are fractions. There are three different forms of fractions:

1. common fractions like $\frac{1}{2}$ and $\frac{3}{4}$,
2. decimal fractions like 0.5 and 0.75, and
3. percents like 50% and 75%.

Each fractional form shows how many parts are in the whole and how many of those parts are named.

With common fractions the denominator shows how many parts are in the whole; the numerator shows how many parts are named. The common fraction $\frac{3}{4}$ shows that there are **4** in the whole and **3** are named.

With decimal fractions the place value of the last digit tells how many parts are in the whole. The digits themselves show how many parts are named. The decimal 0.3 means there are **10** in the whole or group and **3** are named.

With percents, the percent sign means that the whole had been divided into 100 parts. The number in front of the sign shows how many of the parts are named. The amount 25% means there are **100** parts in the whole and **25** of the parts are named.

practice
(See Practice Set DD in the Appendix.)

PART	How many parts in the group?		How many parts are named?		How many parts are not named?	
$\frac{5}{6}$	**a.**	6	**b.**	5	**c.**	1
$\frac{11}{20}$	**d.**	20	**e.**	11	**f.**	9
0.3	**g.**	10	**h.**	3	**i.**	7
0.75	**j.**	100	**k.**	75	**l.**	25
35%	**m.**	100	**n.**	35	**o.**	65
5%	**p.**	100	**q.**	5	**r.**	95

problem set 82

1. What is the product when the sum of 2 and 0.2 is multiplied by the difference of 2 and 0.2?   3.96

2. Arabian camels travel about 3 times as fast as Bactrian camels. If Bactrian camels travel at $1\frac{1}{2}$ miles per hour, then at how many miles per hour do Arabian camels travel? $4\frac{1}{2}$ mph

3. Mark was paid at a rate of $4 per hour for cleaning up a neighbor's yard. If he worked from 1:45 P.M. to 4:45 P.M., how much was he paid? $12

4. Write 57% as a fraction. $\frac{57}{100}$

5. Write $\frac{7}{100}$ as a percent.   7%

6. Write 9% as a fraction. $\frac{9}{100}$

7. What decimal part of a century is 15 years?   0.15

8. Write 0.48 as a reduced common fraction. $\frac{12}{25}$

9. Write $\dfrac{7}{8}$ as a decimal numeral.   0.875

10. $\left(1\dfrac{1}{3} + 1\dfrac{1}{6}\right) - 1\dfrac{2}{3} = \dfrac{5}{6}$ **11.** $1\dfrac{1}{2} \times 3 \times 1\dfrac{1}{9} = 5$

12. $4\dfrac{2}{3} \div 1\dfrac{1}{6} = 4$ **13.** $0.1 + (1 - 0.01) = 1.09$

14. $1.2 \times 0.6 \times 1.4 = 1.008$ **15.** $0.5 \div (1 \div 0.2) = 0.1$

16. Write the standard numeral for $(8 \times 10{,}000) + (4 \times 100) + (2 \times 10)$.   80,420

17. Compare: $1 - 0.2 \bigcirc 1 - \dfrac{1}{2}$ >

18. $6.3 \div 100 = 0.063$

19. What is the greatest common factor of 24 and 32?   8

20. $360 = 6n$ $n = 60$

21. What is the perimeter of the trapezoid?   56 mm

22. One-half of the parallelogram is shaded. What is the area of the shaded part?   12 sq. in.

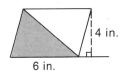

23. One-fourth of the 120 students took wood shop. How many students did not take wood shop?   90

24. $184 - 323 = -139$

25. How many millimeters is 2.5 centimeters?   25 mm

LESSON 83

Changing Percents to Fractions

The three ways of naming parts of a whole are as common fractions, as decimal fractions, and as percents. It is possible to change from one form to another. In this lesson we will practice changing from percents to common fractions. **Percent** means per hundred, so a percent is actually a fraction with a denominator of 100.

$$25\% \text{ means } \frac{25}{100} \qquad 5\% \text{ means } \frac{5}{100}$$

We may write a percent as a fraction simply by writing the number over a denominator of 100, but we should remember to **reduce** the resulting fraction when possible.

$$25\% \text{ means } \frac{25}{100} = \frac{1}{4} \qquad 5\% = \frac{5}{100} = \frac{1}{20}$$

example 83.1   Write 20% as a reduced common fraction.

solution   A percent may be written as a fraction by replacing the percent sign (%) with a denominator of 100. Thus, 20% may be written $\frac{20}{100}$, which reduces to $\frac{1}{5}$.

example 83.2   Write 8% as a fraction.

solution   To change a percent to a fraction, we drop the percent sign and write the number over 100, reducing the resulting fraction when possible.

$$8\% = \frac{8}{100} = \frac{2}{25}$$

practice   Write these percents in fraction form. Reduce when possible.

a. 50% $\frac{1}{2}$ **c.** 75% $\frac{3}{4}$ **e.** 4% $\frac{1}{25}$ **g.** 2% $\frac{1}{50}$

b. 25% $\frac{1}{4}$ **d.** 12% $\frac{3}{25}$ **f.** 6% $\frac{3}{50}$ **h.** 24% $\frac{6}{25}$

problem set 83

1. What is the average of the first four positive odd numbers?   4

2. In 1970 the cost of mailing a letter was 6¢. In 1980 the cost was 15¢. How much **more** did it cost to mail 100 Christmas cards in 1980 than it did in 1970? $9

3. Peter ate seven pumpkins on Friday. If that is only $\frac{1}{3}$ of the world record for pumpkin-eating, what is the world record?   21

*4. Write 20% as a reduced common fraction. $\frac{1}{5}$

*5. Write 8% as a fraction. $\frac{2}{25}$

6. Write $\dfrac{5}{16}$ as a decimal numeral.   0.3125

7. Compare: $1 + \dfrac{3}{4} \bigcirc 1 + 0.7$ >

8. What decimal part of 20 is 3?   0.15

9. What is the area of the parallelogram?   384 sq. cm

10. What is the perimeter of the parallelogram?   82 cm

24 cm
25 cm
16 cm

11. $\left(3\dfrac{1}{8} + 2\dfrac{1}{4}\right) - 1\dfrac{1}{2} = 3\dfrac{7}{8}$

12. $\dfrac{5}{6} \times 2\dfrac{2}{3} \times 3 = 6\dfrac{2}{3}$

13. $5\dfrac{1}{3} \div 4 = 1\dfrac{1}{3}$

14. $(4 - 3.2) \div 10 = 0.08$

15. $1.2 \times 0.11 \times 0.01 =$ 0.00132

16. $8 \div 0.04 = 200$

17. Which digit is in the hundredths' place in 12.345678?   4

18. Round $5\dfrac{1}{8}$ to the nearest whole number.   5

19. Write the prime factorization of 70. $2 \times 5 \times 7$

20. $8m = 4.4$ $m = 0.55$

21. The perimeter of a square is 1 meter. How many centimeters long is each side?   25 cm

22. Forty-five equals $\dfrac{1}{2}$ of what?   90

23. What time is 5 hours and 30 minutes after 9:30 P.M.?
3:00 A.M.

24. $\dfrac{1}{4} - \dfrac{1}{2} =$ (Careful—the answer is less than zero.) $-\dfrac{1}{4}$

25. How long is the line segment?
$\frac{7}{8}$ in.

inches   1   2   3

LESSON 84

Changing Percents to Decimals

Percent means "per hundred" or "hundredths." The number 25% means "twenty-five hundredths," which can be written either as a fraction, $\frac{25}{100}$, or as a decimal, 0.25.

Notice that a percent may be quickly changed to a decimal number by moving the decimal point to the left two places. Study the following changes from percent form to decimal form.

$$35\% = 0.35 \qquad 125\% = 1.25 \qquad 5\% = 0.05$$

example 84.1   Write 15% in decimal form.

solution   The number 15% means $\frac{15}{100}$, which can be written **0.15**. (The decimal form of a percent will have the same digits with the decimal point shifted two places to the left.)

example 84.2   Write 115% as a decimal fraction.

solution   Shifting the decimal point two places to the left, 115% becomes **1.15**.

example 84.3   What decimal number is equal to 1%?

solution   1% is equal to the fraction $\frac{1}{100}$ and to the decimal **0.01**.

practice
(See Practice
Set EE in the
Appendix.)

Write the decimal form of each of these percents.
a. 65% 0.65 **d.** 45% 0.45 **g.** 93% 0.93 **j.** 9% 0.09
b. 225% 2.25 **e.** 375% 3.75 **h.** 105% 1.05
c. 8% 0.08 **f.** 4% 0.04 **i.** 2% 0.02

**problem
set 84**

1. What is the reciprocal of $10\frac{2}{3}$? $\frac{3}{32}$

2. By the time the blizzard was over the temperature had dropped from $17°$ F to $-6°$ F. This was a drop of how many degrees? 23° F

3. The cost to place a telephone call to Tokyo was $3.17 for the first minute plus $1.19 for each additional minute. What was the cost of a 5-minute phone call? $7.93

*4. Write 15% as a decimal fraction. 0.15

*5. Write 115% as a decimal fraction. 1.15

*6. What decimal number is equal to 1%? 0.01

7. Write 60% as a reduced fraction. $\frac{3}{5}$

8. Write 4% as a reduced fraction. $\frac{1}{25}$

9. Write $\frac{13}{20}$ as a decimal numeral. 0.65

10. Compare: $\frac{3}{8}$ ◯ 0.38 <

11. $4\frac{1}{8} - \left(3\frac{1}{4} - 2\frac{1}{2}\right) = 3\frac{3}{8}$ **12.** $\frac{3}{4} \times 2\frac{2}{3} \times \frac{1}{2} = 1$

13. $10 \div 2\frac{1}{2} = 4$ **14.** $6.5 - (4 - 0.32) = 2.82$

15. $6.25 \times 1.6 = 10$ **16.** $0.06 \div 12 = 0.005$

17. Arrange in order of size from least to greatest: $\frac{1}{2}$, 0.4, 30%
30%, 0.4, $\frac{1}{2}$

18. Estimate the product of 89 and 31. 2700

19. $a - 37 = 23$ $a = 60$ **20.** $\dfrac{(300)(60)}{50} = 360$

21. One fourth of the 32 marshmallows burned in the fire. How many did not burn?   24

22. How many square inches is 1 square foot?   144 sq. in.

23. Twenty is one half of what?   40

24. Jim started the 10-kilometer race at 8:22 A.M. He finished the race at 9:09 A.M. How long did it take him to run the race?   47 min.

25. Ten kilometers is how many miles?   6.21 mi.

1 meter	= 1.093 yards
1 kilometer	= 0.621 mile

LESSON 85

Operations with Common Fractions and Decimals

Sometimes there will be common fractions and decimal fractions in the same problem. When fractions and decimals are in the same problem, the problem must be rewritten so that all numbers are written in the same form. They may be written either as fractions or as decimal numbers.

We will look at the problem $0.5 + \frac{1}{4}$. We cannot solve the problem as it is written. We may solve the problem as a decimal problem by changing $\frac{1}{4}$ to 0.25, or we may solve the problem as a fraction problem by changing 0.5 to $\frac{1}{2}$.

As a decimal problem: As a fraction problem:

$$0.5 + \frac{1}{4} =$$ $$0.5 + \frac{1}{4} =$$

$$\left(\frac{1}{4} - 0.25\right)$$ $$\left(0.5 = \frac{1}{2}\right)$$

$$0.5 + 0.25 = \mathbf{0.75}$$ $$\frac{1}{2} + \frac{1}{4} = \frac{2}{4} + \frac{1}{4} = \frac{3}{4}$$

The fraction answer $\frac{3}{4}$ is equal to the decimal 0.75, so either answer is correct. In this book the problems will ask for an answer in one form or the other.

example 85.1 $\frac{1}{2} + 0.3 =$ (decimal answer)

solution Since the problem asks for a decimal answer, we will work the problem in decimal form.

$$\frac{1}{2} + 0.3 =$$

$$0.5 + 0.3 = \mathbf{0.8}$$

practice
(See Practice Set FF in the Appendix.)

Write your answer in decimal form.

a. $\frac{1}{2} + 0.6 = 1.1$

b. $0.3 + \frac{1}{4} = 0.55$

c. $\frac{3}{5} - 0.4 = 0.2$

d. $1.2 \times \frac{3}{4} = 0.9$

e. $3.6 \div \frac{2}{5} = 9$

Write your answer in fraction form.

f. $\frac{3}{4} + 0.5 = 1\frac{1}{4}$

g. $0.8 - \frac{1}{5} = \frac{3}{5}$

h. $0.6 \times \frac{1}{2} = \frac{3}{10}$

i. $\frac{3}{4} \times 0.25 = \frac{3}{16}$

j. $0.4 \div \frac{1}{2} = \frac{4}{5}$

problem set 85

1. What is the quotient when the decimal number ten and six tenths is divided by four hundredths?   265

2. The time in Los Angeles is 3 hours earlier than the time in New York. If it is 1:15 P.M. in New York, what time is it in Los Angeles?   10:15 A.M.

3. Geraldine paid $10 for one dozen photographs costing 75¢ each. How much should she get back in change? $1

Write your answer in the indicated form.

*4. $\frac{1}{2} + 0.3 =$ (decimal)   0.8   **5.** $0.5 + \frac{1}{6} =$ (fraction) $\frac{2}{3}$

6. $0.25 + \frac{3}{5} =$ (decimal)   0.85   **7.** $\frac{3}{5} - 0.4 =$ (fraction) $\frac{1}{5}$

8. $\frac{2}{5} \times 0.12 =$ (decimal)   0.048   **9.** $0.6 \div \frac{3}{4} =$ (fraction) $\frac{4}{5}$

10. Write the decimal numeral for 6%.   0.06

11. $5\frac{1}{2} + 3\frac{7}{8} = 9\frac{3}{8}$ **12.** $3\frac{1}{4} - \frac{5}{8} = 2\frac{5}{8}$ **13.** $\left(4\frac{1}{2}\right)\left(\frac{2}{3}\right) = 3$

14. $3\frac{1}{3} \div 2 = 1\frac{2}{3}$ **15.** $5 \div 1\frac{1}{2} = 3\frac{1}{3}$ **16.** $\frac{5}{6}$ of 30 = 25

17. $4.72 + 12 + 50.4 =$ 67.12   **18.** $10 - 9.87 =$ $0.13

19. Write 7% as a fraction.   **20.** $25 \div 0.08 =$ 312.5
$\frac{7}{100}$

21. What number is next in this sequence? 1, 4, 9, 16, 25, 36, ___   49

22. $3n = 6(4)$ $n =$ 8

23. The perimeter of the rectangle is 48 cm. The width is 6 cm. What is the length?   18 cm

24. $\dfrac{1}{8} - \dfrac{1}{2} = -\dfrac{3}{8}$

25. What was Bonnie's average score on all five tests?   92%

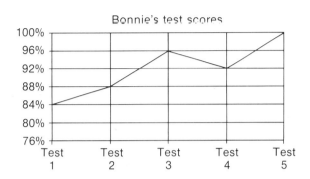

Bonnie's test scores

LESSON 86

Finding Missing Factors

We have solved problems like the following in which the unknown factor was a whole number.

$$5n = 20, \qquad n =$$

In this lesson we will begin to solve problems in which the unknown factor is a decimal fraction or common fraction. To solve these problems we will continue to use the following rule.

> **We can find an unknown factor by dividing the product by the known factor.**

example 86.1 $5n = 21 \qquad n =$

solution To find an unknown factor, we divide the product by the known factor. Note: We will end our division with fractions unless there are decimal numbers in the problem.

$$\begin{array}{r} 4\frac{1}{5} \\ 5{\overline{\smash{\big)}\,21}} \\ \underline{20} \\ 1 \end{array}$$

$$n = 4\frac{1}{5}$$

example 86.2 $0.6m = 0.048$ $m =$

> *solution*   Again we will find the unknown factor by dividing the product by the known factor. This time we will write our answer as a decimal fraction.

$$\begin{array}{r} 0.08 \\ 0.6\overline{)0.0.48} \\ 48 \\ \hline 0 \end{array}$$

$$m = \mathbf{0.08}$$

example 86.3 $45 = 4x$ $x =$

> *solution*   An equals sign is not directional. It simply states that two amounts are equal. In this case the product is 45 and the known factor is 4. We divide 45 by 4 to find the unknown factor.

$$\begin{array}{r} 11\frac{1}{4} \\ 4\overline{)45} \\ 4 \\ \hline 5 \\ 4 \\ \hline 1 \end{array}$$

$$x = \mathbf{11\frac{1}{4}}$$

practice
a. $12y = 400$ $y = 33\frac{1}{3}$ **d.** $0.3t = 2.34$ $t = 7.8$

b. $6w = 21$ $w = 3\frac{1}{2}$ **e.** $8s = 32.4$ $s = 4.05$

c. $500 = 3f$ $f = 166\frac{2}{3}$ **f.** $0.36 = 5n$ $n = 0.072$

problem set 86

1. If the divisor is 12 and the quotient is 24, what is the dividend?   288

2. The brachiosaurus, one of the largest dinosaurs, weighed only $\frac{1}{4}$ as much as a blue whale. The blue whale weighs 140 tons when full grown. What was the weight of the brachiosaurus?   35 tons

3. Buddha lived from approximately 563 B.C. to 483 B.C. About how old was he when he died?   80

4. $32m = 1504$ $m = 47$

*5. $5n = 21$ $n = 4\frac{1}{5}$

6. $6n = 25$ $n = 4\frac{1}{6}$

7. $6.5 + \dfrac{3}{5} =$ (decimal answer) 7.1

8. $3\dfrac{1}{3} + 0.25 =$ (fraction answer) $3\dfrac{7}{12}$

9. Write 175% as a decimal numeral. 1.75

10. Write 65% as a fraction. (Always reduce when possible!)
$\frac{13}{20}$

11. $12\dfrac{1}{5} - 3\dfrac{4}{5} = 8\dfrac{2}{5}$ **12.** $6\dfrac{2}{3} \times 1\dfrac{1}{5} = 8$ **13.** $11\dfrac{1}{2} \div 2 = 5\dfrac{3}{4}$

14. $4.75 + 12.6 + 10 = 27.35$ **15.** $0.35 \div 4 = 0.0875$

16. Divide: $\dfrac{45}{1.2} = 37.5$

17. Write the decimal numeral twelve and five hundredths.
12.05

18. $35 - (0.35 \times 100) = 0$

19. What is the least common multiple of 2, 3, and 4? 12

20. If $a = 15$, then $2a - 5$ equals what? 25

21. What is the area of the parallelogram?
450 sq. mm

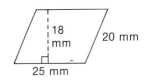

22. "All rectangles are parallelograms." (True or false?)
true

23. Charles spent $\frac{1}{10}$ of his 100 shillings. How many shillings did he still have? 90

24. The temperature rose from $-18°$ F to $19°$ F. How many degrees did the temperature increase? 37° F

25. How many **centimeters** long is the line segment? 4.5 cm

LESSON 87

Area of a Triangle

 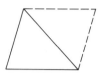

Notice that the area of any triangle is half the area of a **parallelogram**. We have seen that the area of a parallelogram can be found by multiplying the base times the height ($A = bh$), so the area of a triangle can be determined by finding **half** of the area of the parallelogram of which it is a part. It is not necessary to draw the parallelogram each time. Multiplying the **base** of the triangle by the **height** of the triangle gives the area of the parallelogram. Dividing that amount by two gives the area of the triangle. The formula for finding the area of a triangle is $A = \frac{1}{2}bh$, which may also be written $A = \frac{bh}{2}$.

example 87.1   Find the area of the triangle.

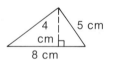

solution   The area of a triangle is half the product of the base and height. The height must be **perpendicular** to the base. The height in this case is 4 cm. Half the product of 8 cm and 4 cm is

$$\text{Area} = \frac{8 \times 4}{2} = \textbf{16 sq. cm}$$

example 87.2   Find the area of the triangle.

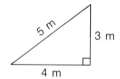

solution   We find the area by multiplying the base and the height, then dividing by 2. With a right triangle we use the **perpendicular** sides as the base and height.

$$\text{Area} = \frac{3 \times 4}{2} = \textbf{6 sq. m}$$

practice Find the area of each triangle.

a.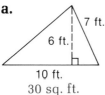
10 ft.
30 sq. ft.

b.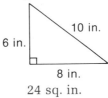
24 sq. in.

c.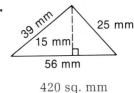
420 sq. mm

problem set 87

1. Write the numeral five million, eight hundred seventy thousand, eighty-six.   5,870,086

2. If Pinocchio's nose grows $\frac{1}{4}$ inch per lie, then how many lies has he told if his nose has grown 4 inches?   16

3. Mark wants to buy a new baseball glove costing $50. He has $14 and earns $4 per hour cleaning yards. How many hours must he work to have enough money to buy the glove?   9 hr.

*4. Find the area of the triangle.
16 sq. cm

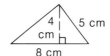

5. $42w = 3192$ $w = 76$

6. $9\ m = 20$ $m = 2\frac{2}{9}$

*7. Find the area of the triangle.
6 sq. m

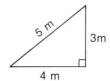

8. Write 5% as a decimal numeral.   0.05

9. $2\frac{1}{5} - 0.6 =$ (fraction answer) $1\frac{3}{5}$

10. $2.5 \times \frac{2}{5} =$ (decimal answer)   1

11. $\frac{2}{3} \div 0.25 =$ (fraction answer) $2\frac{2}{3}$

12. $\frac{2}{3} + \frac{3}{2} = 2\frac{1}{6}$ **13.** $(5)\left(1\frac{3}{5}\right) = 8$ **14.** $5 \div 1\frac{3}{5} = 3\frac{1}{8}$

15. $25 \times 2.5 \times 0.01 = $ 0.625   **16.** $0.025 \times 100 = $ 2.5

17. $\dfrac{12}{0.25} = $ 48

18. What is the sum of the first eight numbers of this sequence?
1, 3, 5, 7, . . . 64

19. Which of these is a composite number? 61, 71, 81, 101
81

20. Round the decimal number one and twenty-three hundredths to the nearest tenth. 1.2

21. Albert baked 5 dozen cookies and ate $\frac{7}{12}$ of them. How many cookies are left? 25

22. $75 - 750 = $ −675   **23.** $76 \cdot 1 = 76 + \boxed{}$ 0

24. How many millimeters is 4 meters? 4000 mm

25. How long is the line segment?
$2\frac{1}{4}$ in.

LESSON 88

Comparing Negative Numbers

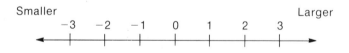

We know that 2 is less than 3, but how would we compare −2 and −3? Looking at the number line, we see that as we move further and further to the left, the numbers get smaller and smaller. Since −3 is to the left of −2, it is less than −2. This makes sense because a number which is 3 less than zero is less than a number which is 2 less than zero.

example 88.1 Compare: $-5 \bigcirc -2$

solution   A number which is 5 less than zero is less than a number which is 2 less than zero. **−5 < 2.**

example 88.2   Arrange in order from least to greatest: 0, 2, −3

solution   Negative numbers are less than zero, so −3 is less than 0, and 0 is less than 2. We write the answer in this order: **−3, 0, 2**

practice   Compare.   Arrange in order from least to greatest.
    a. −3 ◯ −5   >   **e.** −1, −2, 0   −2, −1, 0
    b. 3 ◯ −2   >   **f.** 3, −1, −2   −2, −1, 3
    c. −1 ◯ 0.2   <   **g.** $\dfrac{1}{2}$, −1, 0   −1, 0, $\dfrac{1}{2}$
    d. 3 − 2 ◯ 2 − 3
        >   **h.** 1, −1, 0.1   −1, .1, 1

problem   **1.** What is the difference when the product of $\frac{1}{2}$ and $\frac{1}{2}$ is sub-
set 88   tracted from the sum of $\frac{1}{4}$ and $\frac{1}{4}$? $\frac{1}{4}$

    2. A dairy cow can give 4 gallons of milk per day. How many cups of milk is that? (1 gallon = 4 quarts, 1 quart = 4 cups.)
       64 cups

    3. The recipe called for $\frac{3}{4}$ cup of sugar. If the recipe is doubled, how much sugar should be used? $1\frac{1}{2}$ cups

    *4. Compare: −5 ◯ −2   <

    *5. Arrange in order of size from least to greatest: 0, 2, −3
       −3, 0, 2

    6. $7n = 30$ $n = 4\frac{2}{7}$

    7. What is the area of the triangle?
       20 sq. in.

    8. What is the area of the parallelogram?
       40 sq. in.

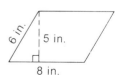

    9. $5.3 + \dfrac{7}{20} =$ (decimal answer)   5.65

10. $(0.6)\left(\dfrac{3}{4}\right) =$ (decimal answer) 0.45

11. $3\dfrac{1}{4} - 0.5 =$ (fraction answer) $2\dfrac{3}{4}$

12. $0.8 \div \dfrac{2}{3} =$ (fraction answer) $1\dfrac{1}{5}$

13. $\$6 + 6¢ + \$6.66 =$ $12.72 **14.** $\$0.07 \times 100 =$ $7

15. $\$6 \div 8 =$ $0.75 **16.** $1\dfrac{3}{5} \times 10 \times \dfrac{1}{4} = 4$

17. $7\dfrac{1}{2} \div 3 = 2\dfrac{1}{2}$ **18.** $3 \div 7\dfrac{1}{2} = \dfrac{2}{5}$

19. What is the place value of the 7 in 987,654.321? thousands

20. Write the decimal numeral five hundred ten and five hundredths. 510.05

21. $30 + 60 + m = 180$ $m = 90$

22. Round the product of 2.4 and 2.6 to the nearest tenth. 6.2

23. Half of the students are girls. Half of the girls have brown hair. Half of the brown-haired girls wear their hair long. Of the 32 students, how many are long-brown-haired girls? 4

24. Ninety is $\dfrac{1}{2}$ of what? 180

25. How many more books has Mary read than Pat? 6

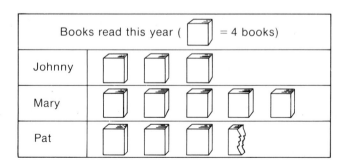

LESSON
89

Naming Solids

PRISMS:	Triangular Prism	
	Rectangular Prism	
	Cube	
Pyramids		
Cylinder		
Cone		
Sphere		

The table on this page lists some three-dimensional shapes. Three-dimensional shapes are shapes that take up space. Three-dimensional shapes are often called **solids**. You should be able to recognize, name, and draw each of these shapes. Notice that when these shapes are drawn, the edges which are hidden from the viewer can be indicated by using dotted lines.

Prisms are solids with a polygon of a constant size "running through" the shape. A triangular prism has a triangle running through the shape. A rectangular prism has a rectangle running through the shape. A cube is a special type of rectangular prism with a square running through the shape.

example 89.1   Name this shape.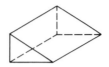

solution   This shape has a triangle running through it. It is a **triangular prism**.

example 89.2   Draw a cube.

solution   Cubes and other prisms can be drawn by drawing the polygon which runs through it twice, like this. Then draw lines to connect the "corners" and use dotted lines to draw the edges that are hidden from view.

problem set 89

1. What is the average of 4.2, 4.8, and 5.1?   4.7

2. The movie is 127 minutes long. If it begins at 7:15 P.M., when will it be over?   9:22 P.M.

3. Silvia was thinking of a number less than 90 which one can get to by counting by sixes or counting by fives, but not by counting by fours. Of what number was she thinking?   30

*4. Name this shape.   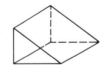   triangular prism

*5. Draw a cube.

6. Name this shape.   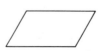   parallelogram

7. Compare: $\dfrac{5}{8}$ ◯ 0.58   >

8. Arrange in order of size from least to greatest: $-2, 3, -4$
$-4, -2, 3$

9. $25n = 700$ $n = 28$

10. What is the area of the triangle?
150 sq. mm

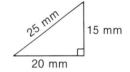
25 mm · 15 mm · 20 mm

11. $6.25 - \dfrac{5}{8} =$ (decimal answer)   5.625

12. Write 125% as a decimal numeral.   1.25

13. Write 28% as a reduced fraction. $\dfrac{7}{25}$

14. $(0.49)(0.51) =$ 0.2499   **15.** $0.625 \div 10 =$ 0.0625   **16.** $\dfrac{25}{0.8} = 31.25$

17. $3\dfrac{3}{8} + 3\dfrac{3}{4} = 7\dfrac{1}{8}$ **18.** $5\dfrac{1}{8} - 1\dfrac{7}{8} = 3\dfrac{1}{4}$ **19.** $6\dfrac{2}{3} \times \dfrac{3}{10} \times 4 = 8$

20. One third of the two dozen knights were on horseback. How many knights were not on horseback?   16

21. What is the average of $6.30, 57¢, and $5.31? $4.06

22. What percent of a century is a decade?   10%

23. The shape at right is made up of how many cubes?   12

24. Round the sum of forty-eight hundredths and eighty-four hundredths to the nearest tenth.   1.3

25. How long is the line segment? $2\dfrac{3}{8}$ in.

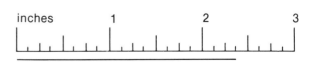

LESSON 90

Faces, Edges, Vertices

We name the parts of a solid **faces**, **edges**, and **vertices** as illustrated below.

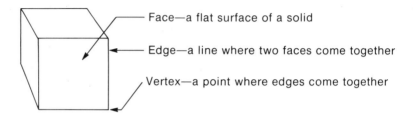

Face—a flat surface of a solid

Edge—a line where two faces come together

Vertex—a point where edges come together

(The plural of **vertex** is either **vertices** or **vertexes**.)

practice   Answering these questions will help to fix these ideas in our minds.   6 (square)
a. A cube has how many faces? (What shape is each face?)
b. How many edges does a cube have?   12
c. How many vertices does a cube have?   8

Here is a pyramid with a square base.
One face is a square; the rest are triangles.
d. How many faces are there in all?   5
e. How many edges are there?   8
f. How many vertices are there?   5

Here is a triangular prism.
g. How many faces does it have?   5
h. How many edges?   9
i. How many vertices?   6
j. A rectangular prism has how many more edges than vertices?   4
k. A rectangular prism has how many more faces than a triangular prism?   1

problem set 90

1. What is the reciprocal of $12\frac{1}{2}$? $\frac{2}{25}$

2. Sunrise was at 6:15 A.M., and sunset was at 5:45 P.M. How many minutes were there from sunrise to sunset?
690 min.

3. What number is equal to all of your fingers plus half your toes, minus your knees and elbows? 11

4. Name this shape. 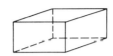 rectangular prism

5. Name this shape.   trapezoid

6. This is a pyramid with a triangular base.
   a. How many faces does it have? 4
   b. How many edges does it have? 6
   c. How many vertexes does it have? 4

7. What is the perimeter of the parallelogram? 4.4 in.

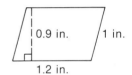

8. What is the area of the parallelogram?
   1.08 sq. in.

9. Write 225% as a decimal numeral. 2.25

10. Write 64% as a reduced fraction. $\frac{16}{25}$

11. $6\frac{2}{3} + 1\frac{3}{4} = 8\frac{5}{12}$ 12. $5 - 1\frac{2}{5} = 3\frac{3}{5}$ 13. $4\frac{1}{4} - 3\frac{5}{8} = \frac{5}{8}$

14. $3 \times \frac{3}{4} \times 2\frac{2}{3} = 6$ 15. $2\frac{1}{4} \div 3 = \frac{3}{4}$ 16. $2\frac{1}{2} \div 3\frac{3}{4} = \frac{2}{3}$

17. $8.4 + \frac{3}{4} =$ (decimal answer) 9.15

18. What decimal part is **not** shaded? 0.875

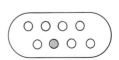

19. If $\frac{5}{6}$ of the 300 seeds sprouted, how many seeds did **not** sprout? 50

20. $6y = 10$ $y = 1\frac{2}{3}$ 21. $2.1 - 3 = -0.9$

22. What is the area of the triangle?
12 sq. ft.

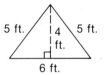

23. Round $6.5432 to the nearest cent. $6.54

24. What time is 2 hours and 15 minutes before 1:05 P.M.?
10:50 A.M.

25. AC is 56 mm. BC is 26 mm. How long is AB?   30 mm

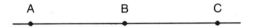

LESSON
91

Exponents

An **exponent** is a number written above and to the right of another number. An exponent is used to indicate repeated multiplication. The exponent shows how many times the other number is to be used as a factor. The number used as a factor is called the **base**.

In the number 3^2, 2 is the exponent and 3 is the base. We see that the expression tells us to take two 3's and multiply them together: $3^2 = 3 \times 3 = 9$.

In 2^3 the exponent is 3, and it shows that the base 2 is to be used as a factor 3 times: $2^3 = 2 \times 2 \times 2 = 8$.

examples and solutions

$$3^4 = 3 \times 3 \times 3 \times 3 = \mathbf{81} \qquad 10^3 = 10 \cdot 10 \cdot 10 = \mathbf{1000}$$

$$5^2 = 5(5) = \mathbf{25} \qquad\qquad 2^5 = (2)(2)(2)(2)(2) = \mathbf{32}$$

We read numbers with exponents as **powers**.
We read 3^4 as "three to the fourth power."
We read 5^2 as "five to the second power" or "five squared."
We read 10^3 as "ten to the third power" or "ten cubed."
We read 2^5 as "two to the fifth power."

We commonly read the exponents 3 and 2 as "cubed" and "squared" since a number used as a factor 3 times gives the volume of a cube, and a number used as a factor 2 times gives the area of a square.

practice

a. $5^3 = 125$

b. $4^2 = 16$

c. $10^4 = 10{,}000$

d. $2^6 = 64$

e. $12^2 = 144$

f. What does two to the fourth power equal? 16

g. What is the value of six cubed? 216

h. How much is ten squared? 100

i. $(0.1)^2 = 0.01$

j. $(1.1)^2 = 1.21$

problem set 91

1. Change the mixed number $6\frac{2}{3}$ to an improper fraction. $\frac{20}{3}$

2. Twelve out of fifty is the same thing as how many out of one hundred? 24

3. If the average of three numbers is 144, then what is the sum of the three numbers? 432

4. Is it true or false that all quadrilaterals are polygons? true

*5. $3^4 = 81$

*6. $10^3 = 1000$

7. $7^2 = 49$

8. Arrange in order of size from least to greatest: 0.1, 1, −1
$-1, 0.1, 1$

9. If $\frac{5}{6}$ of the 30 members were present, then how many were absent? 5

10. What is 0.6 of 30? 18

11. $\dfrac{(24)(36)}{48} = 18$

12. $12\frac{5}{6} + 15\frac{1}{3} = 28\frac{1}{6}$

13. $100 - 9.9 = 90.1$

14. $6\frac{2}{3} \times 1\frac{1}{5} = 8$

15. $1 \div 3\frac{3}{4} = \frac{4}{15}$

16. $0.23 \times 4.06 = 0.9338$

17. $7.4 - 3\frac{3}{4} =$ (decimal answer) 3.65

18. $16 + 12\frac{3}{10} + 8.4 =$ (fraction answer) $36\frac{7}{10}$

19. Write 8% as a decimal.   0.08

20. Write 80% as a reduced fraction. $\frac{4}{5}$

21. Estimate the product of 6.95 and 12.1 to the nearest whole number.   84

22. 5n = 345   n = 69

23. What is the area of the triangle?
24 sq. cm

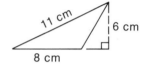

24. This shape has six rectangular faces. What is the name of this shape?
rectangular prism

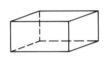

25. How long is the line segment?
$1\frac{1}{8}$ in.

LESSON
92

Writing Mixed Numbers as Decimals

A **mixed number** is a whole number and a fraction written together, like $1\frac{1}{2}$. Mixed numbers may be written in decimal form just as fractions are written in decimal form. The whole number part of a decimal number is written in front of (to the left of) the decimal point. The fraction part is behind (to the right of) the decimal point. When writing a mixed number as a decimal numeral, we write the whole number part in front of the decimal point; then we change the fraction to a decimal and write it after the whole number part.

$$1\ \ \frac{1}{2}$$
$$1\ \ 5$$

example 92.1 Write $2\frac{1}{4}$ as a decimal numeral.

solution The whole number part is in front of the decimal point (2.). The fraction $\frac{1}{4}$ becomes the decimal 0.25. When we put them together, we get **2.25**.

example 92.2 $3.5 + 3\frac{1}{5} =$ (decimal answer)

solution Both numbers must be written in the same form to add. We add 3.5 and 3.2 and get **6.7**.

$$3.5 + 3\frac{1}{5}$$
$$3.5 + 3.2 = 6.7$$

practice Write these mixed numbers in decimal form.

a. $1\frac{1}{4}$ 1.25   **d.** $3\frac{1}{10}$ 3.1   **g.** $6\frac{3}{4}$ 6.75   **j.** $20\frac{1}{20}$ 20.05

b. $2\frac{1}{2}$ 2.5   **e.** $4\frac{3}{5}$ 4.6   **h.** $12\frac{4}{5}$ 12.8   **k.** $1\frac{3}{25}$ 1.12

c. $3\frac{2}{5}$ 3.4   **f.** $5\frac{1}{8}$ 5.125   **i.** $8\frac{3}{8}$ 8.375   **l.** $16\frac{17}{100}$ 16.17

problem set 92

1. What is the difference of ten million, fifty-five thousand, two hundred one and six million, seven hundred eight thousand, four hundred eighty?   3,346,721

2. What percent of a century is a decade?   10%

3. If the sum of three numbers is 144, then what is the average of the three numbers?   48

4. How many smaller cubes have been put together to form the larger cube? 8

*5. Write $2\frac{1}{4}$ as a decimal numeral.   2.25

*6. $3.5 + 3\frac{1}{5} =$ (decimal answer)   6.7

7. $5.4 - 3\dfrac{3}{4} =$ (decimal answer)   1.65

8. $3^3 =$ 27   **9.** $24 \cdot 25 =$ 600   **10.** $\dfrac{1000 - 8}{16} =$ 62

11. Twenty of the two dozen members voted yes. What fraction of the members voted yes? $\frac{5}{6}$

12. $7 + 8\dfrac{3}{4} + 3\dfrac{5}{8} =$ **13.** $9\dfrac{1}{3} - 4\dfrac{3}{4} = 4\dfrac{7}{12}$ **14.** $4\dfrac{1}{2} \div 9 = \dfrac{1}{2}$
$19\frac{3}{8}$

15. $6.75 + 12 + 4.6 =$ 23.35   **16.** $1 \div 0.025 =$ 40

17. $(5.75)(8.4) =$ 48.3

18. A triangular prism has how many faces?   5

19. Fifteen centimeters is what percent of a meter?   15%

20. Write 85% as a reduced fraction. $\frac{17}{20}$

21. Round the decimal number one hundred twenty-five thousandths to the nearest tenth.   0.1

22. $6.2 + x = 10$ $x =$ 3.8

23. What is the area of one of the triangles?   234 sq. mm

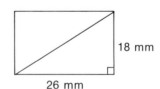

18 mm

26 mm

24. What is the missing dividend? $6\overline{)}$?   90   (quotient 15)

25. To what decimal number is the arrow pointing?   1.4

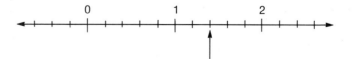

LESSON
93

Writing Fractions as Percents, Part 1

The number 100% means $\frac{100}{100}$, which equals 1. We remember that we can change the name of a fraction without changing its value by multiplying by a different name for 1. The percent name for 1 is 100%. We may change any fraction to a percent by multiplying by 100%. This is another use of the "Name-Changer Machine."

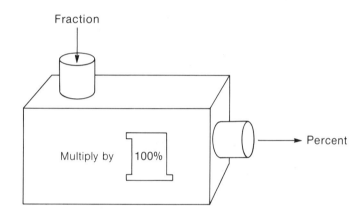

Rule: To change a number to a percent, multiply by 100%.

example 93.1 Change $\frac{1}{2}$ to percent.

solution **To change a number to percent, we multiply by 100%.**

$$\frac{1}{2} \times \frac{100\%}{1} = \frac{100\%}{2} = \mathbf{50\%}$$

example 93.2 Change the fraction $\frac{3}{5}$ to its percent form.

solution **To change a number to percent, we multiply by 100%.**

$$\frac{3}{5} \times \frac{100\%}{1} = \frac{300\%}{5} = \mathbf{60\%}$$

practice   Change each fraction below to its percent equivalent.

a. $\dfrac{1}{2} = 50\%$ d. $\dfrac{1}{5} = 20\%$ g. $\dfrac{4}{5} = 80\%$ j. $\dfrac{1}{20} = 5\%$

b. $\dfrac{1}{4} = 25\%$ e. $\dfrac{2}{5} = 40\%$ h. $\dfrac{3}{10} = 30\%$

c. $\dfrac{3}{4} = 75\%$ f. $\dfrac{3}{5} = 60\%$ i. $\dfrac{1}{25} = 4\%$

problem set 93

1. When the product of $\frac{1}{2}$ and $\frac{1}{3}$ is subtracted from the sum of $\frac{1}{2}$ and $\frac{1}{3}$, what is the difference? $\frac{2}{3}$

2. On the Celsius scale water freezes at $0°$ C and boils at $100°$ C. What temperature is halfway between the freezing and boiling temperatures?   50° C

3. If AB is $\frac{1}{3}$ of AC, and if AC is 12 cm, then how long is BC?
   8 cm

4. What decimal part of the circles is not shaded?   0.6

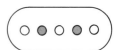

*5. Change the fraction $\dfrac{3}{5}$ to its percent form.   60%

6. What is the percent equivalent of $\dfrac{7}{10}$?   70%

7. $6.4 - 6\dfrac{1}{4} =$ (decimal answer)   0.15

8. $10^4 = 10{,}000$ 9. $4 \cdot 12 = 3n$ $n = 16$

10. How much is $\dfrac{3}{4}$ of 360?   270   11. How much is 0.3 of 360?   108

12. $3\dfrac{1}{2} + 1\dfrac{3}{4} + 4\dfrac{5}{8} = 9\dfrac{7}{8}$ 13. $\dfrac{9}{10} \times \dfrac{5}{6} \times \dfrac{8}{9} = \dfrac{2}{3}$

14. How many halves are in three fourths? $1\frac{1}{2}$

15. $8.47 + 95¢ + $12 = $21.42   **16.** 36.45 ÷ 100 = 0.3645

17. (1 + 0.99) × (1 − 0.99) − 0.0199

18. $3.3 - 2\frac{2}{5} =$ (fraction answer) $\frac{9}{10}$

19. Ninety percent were correct. What percent were incorrect?
10%

20. Write the decimal numeral one hundred twenty and three hundredths.   120.03

21. Arrange in order of size from **greatest** to **least:** $\frac{2}{5}, \frac{5}{2}, -2.5$
$\frac{5}{2}, \frac{2}{5}, -2.5$

22. A pyramid with a square base has how many edges?   8

23. What is the area of one of the triangles in this parallelogram?
40 sq. in.

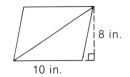
8 in.
10 in.

24. On the coldest day of the year the temperature was −37° F. On the hottest day it was 103° F. What was the range of temperature for the year?   140° F

25. How many millimeters long is the line segment?   30 mm

cm   2   4

LESSON
94

Writing Fractions as Percents, Part 2

In the last lesson we learned a rule for changing numbers into percents: "To change a number to a percent we multiply by 100%." The fractions we have changed to percents so far have equaled whole percents. In this lesson we will change fractions into percents which end with a fraction. To change the fraction $\frac{1}{7}$ to a percent, we multiply by 100%.

$$\frac{1}{7} \times \frac{100\%}{1} = \frac{100\%}{7}$$

The division is done the way we show here. We will end uneven divisions with a fraction.

$$\begin{array}{r} 14\frac{2}{7}\% \\ 7\overline{)100\%} \\ \underline{7} \\ 30 \\ \underline{28} \\ 2 \end{array}$$

example 94.1   Change $\dfrac{2}{3}$ to its percent equivalent.

solution   To change a fraction to a percent, we multiply by 100%.

$$\frac{2}{3} \times \frac{100\%}{1} = \frac{200\%}{3} = \mathbf{66\frac{2}{3}\%}$$

practice
(See Practice
Set GG in the
Appendix.)

Change each of these fractions to its percent equivalent. End uneven divisions with reduced fractions.

a. $\dfrac{3}{50} = 6\%$ **d.** $\dfrac{1}{100} = 1\%$ **g.** $\dfrac{6}{5} = 120\%$ **j.** $\dfrac{11}{4} = 275\%$

b. $\dfrac{1}{3} = 33\frac{1}{3}\%$ **e.** $\dfrac{1}{6} = 16\frac{2}{3}\%$ **h.** $\dfrac{5}{6} = 83\frac{1}{3}\%$ **k.** $\dfrac{4}{11} = 36\frac{4}{11}\%$

c. $\dfrac{1}{8} = 12\frac{1}{2}\%$ **f.** $\dfrac{19}{20} = 95\%$ **i.** $\dfrac{1}{9} = 11\frac{1}{9}\%$ **l.** $\dfrac{5}{12} = 41\frac{2}{3}\%$

**problem
set 94**

1. If school begins at 8:00 A.M. and ends at 3:10 P.M., then how many minutes long is the school day?   430 min.

2. Jeff is 1.67 meters tall. How many centimeters tall is Jeff?
167 cm

3. If $\frac{5}{8}$ of the 40 seeds sprouted, then how many seeds did not sprout?   15

4. Change this amount from expanded notation to standard notation: $(5 \times 100) + (6 \times 10) + (7 \times \frac{1}{10}) + (3 \times \frac{1}{100})$
560.73

*5. Change $\dfrac{2}{3}$ to its percent equivalent. $66\frac{2}{3}\%$

6. Write $\dfrac{3}{2}$ in percent form.   150%

7. What percent is equal to the fraction $\dfrac{3}{8}$? $37\frac{1}{2}\%$

8. Ten percent is the same as what fraction? $\dfrac{1}{10}$

9. $25^2 = $ 625

10. What is 0.03 of 3000?   90

11. How many cubes make this shape?
36

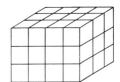

12. $\dfrac{3}{4} + \dfrac{3}{5} = 1\frac{7}{20}$ **13.** $18\dfrac{1}{8} - 12\dfrac{1}{2} = 5\frac{5}{8}$ **14.** $3\dfrac{3}{4} \times 2\dfrac{2}{3} \times 1\dfrac{1}{10} =$ 11

15. How many fourths are in $2\dfrac{1}{2}$?   10

16. $12 + 8.75 + 6.8 = $ 27.55   **17.** $0.375 \times 0.16 = $ 0.06

18. $6\dfrac{2}{5} \div 0.016 = $ (decimal form)   400

19. Estimate the sum of $6\frac{1}{4}$, 4.95, and 8.21 to the nearest whole number.   19

20. Round three and four hundred fifty-six thousandths to the nearest hundredth.   3.46

21. Arrange in order from least to greatest: $\dfrac{1}{4}$, 4%, 0.4
$4\%, \frac{1}{4}, 0.4$

22. y + 3.4 = 5   y = 1.6

23. What is the dividend? 8) ?   96

$$\overset{12}{8\overline{)}}$$

24. A cube has edges which are 6 cm long. What is the area of each face of the cube?   36 sq. cm

25. AB = 24 mm. AC = 42 mm. How long is BC?   18 mm

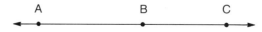

A   B   C

LESSON 95

Writing Decimals as Percents

We have seen that fractions can be named as percents by multiplying by the name of 1 called 100%. Decimal fractions may be named as percents in the same way.

> **To change any number to a percent, multiply by 100%.**

To change 0.4 to a percent, multiply by 100%:

$$0.4 \times 100\% = 40\% \qquad \begin{array}{r} 100\% \\ \times\ 0.4 \\ \hline 40.0\% \end{array}$$

Remember the shortcut for multiplying by 10, 100, 1000, or any other power of 10: The digits stay the same but the place values change. When multiplying by 100 we can change the place values by shifting the decimal point two places to the right. Thus, when we multiply 0.4 by 100 percent, we shift the decimal point two places to the right.

$$0.4 \times 100\% = 0.\underset{\frown}{40}.\% = 40\%$$

example 95.1   Write 0.12 as a percent.

solution To change a number to a percent, we multiply by 100%. The quick way to multiply by 100% is to shift the decimal point to the right two places. Changing a decimal to a percent always shifts the decimal point two places to the right.

$$0.12 \times 100\% = 0.\underset{\curvearrowright}{12}\% = \mathbf{12\%}$$

practice
(See Practice
Set HH in the
Appendix.)

Change these decimal fractions to percents by multiplying by 100%.

a. 0.5 50% **f.** 0.06 6% **k.** 0.9 90%
b. 0.15 15% **g.** 1.2 120% **l.** 1 100%
c. 0.04 4% **h.** 1.25 125% **m.** 2.3 230%
d. 0.8 80% **i.** 1.04 104% **n.** 4 400%
e. 0.24 24% **j.** 0.09 9% **o.** 0.1 10%

problem set 95

1. When the sum of 2.2 and 1.9 is subtracted from the product of 2.2 and 1.9, what is the difference? 0.08

2. An object weighing 5.04 kilograms weighs the same as how many objects each weighing 0.42 kilogram? 12

3. If the average of 8 numbers is 12, then what is the sum of the 8 numbers? 96

4. What is the name of a quadrilateral which has one pair of sides which are parallel and one pair which are not parallel? trapezoid

*5. Write 0.12 as a percent. 12%

6. Write 1.2 as a percent. 120%

7. Write $\frac{5}{8}$ as a percent. $62\frac{1}{2}\%$

8. Three of these are equal. Which one is different? 1, 100%, 0.1, $\frac{10}{10}$ 0.1

9. $11^3 = 1331$

10. How much is $\frac{5}{6}$ of 360? 300

11. $\dfrac{(45)(54)}{81} = $ 30   **12.** $\dfrac{30}{0.08} = $ 375   **13.** $\dfrac{\frac{1}{2}}{\frac{1}{3}}$ $1\frac{1}{2}$

14. $2\dfrac{1}{2} + 3\dfrac{1}{3} + 4\dfrac{1}{6} = $ 10   **15.** $6 \times 5\dfrac{1}{3} \times \dfrac{3}{8} = $ 12

16. $\$1.25 \times 100 = $ $125   **17.** What is 0.12 of 6.4?   0.768

18. $5.3 - 3\dfrac{3}{4} = $ (decimal answer)   1.55

19. One half of the students were absent. What percent were present?   50%

20. Which digit in 6.857 has the same place value as the 3 in 573?   6

21. Estimate the sum of 0.635 and 1.497 to the nearest tenth.
2.1

22. $3n = 4.2 + 1.5$ $n = $ 1.9   **23.** $\dfrac{1}{3} - 1 = -\dfrac{2}{3}$

24. The perimeter of the rectangle is 100 mm. What is its length?
29 mm

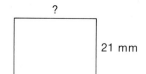

25. How long is the line segment?
$1\frac{3}{4}$ in.

LESSON
96

Writing Mixed Numbers as Percents

Mixed numbers can be written as percents greater than 100%. The number 1 is equal to 100%, so numbers greater than 1 equal percents greater than 100%.

We change fractions to percents by multiplying by 100%. We also change mixed numbers to percents by multiplying by 100%.

example 96.1 Write $2\frac{1}{4}$ as a percent.

solution We will show two methods.

Method 1: We change the mixed number to fraction form before multiplying.

$$2\frac{1}{4} = \frac{9}{4}$$

$$= \frac{9}{4} \times \frac{100\%}{1} = \frac{900\%}{4} = \mathbf{225\%}$$

Method 2: We may shorten the work somewhat by changing the whole number to a percent and the fraction to a percent. Then we combine them.

$$2 \quad \frac{1}{4}$$

$$200\% + 25\% = \mathbf{225\%}$$

practice Change each mixed number into percent form.

a. $1\frac{3}{4}$ 175%   **e.** $5\frac{3}{5}$ 560%   **i.** $9\frac{2}{3}$ $966\frac{2}{3}\%$ **m.** $13\frac{1}{6}$ $1316\frac{2}{3}\%$

b. $2\frac{1}{2}$ 250%   **f.** $6\frac{1}{10}$ 610%   **j.** $10\frac{3}{10}$ 1030%   **n.** $14\frac{1}{7}$ $1414\frac{2}{7}\%$

c. $3\frac{1}{5}$ 320%   **g.** $7\frac{3}{20}$ 715%   **k.** $11\frac{1}{9}$ $1111\frac{1}{9}\%$

d. $4\frac{1}{4}$ 425%   **h.** $8\frac{1}{3}$ $833\frac{1}{3}\%$ **l.** $12\frac{1}{8}$ $1212\frac{1}{2}\%$

problem set 96 **1.** How many quarter-pound hamburgers can be made from 100 pounds of ground beef?   400

2. On the Fahrenheit scale, water freezes at 32° F and boils at 212° F. What temperature is halfway between the freezing and boiling temperatures?   122° F

3. If the sum of 8 numbers is 48, then the average of the 8 numbers is what?   6

4. Compare: $\dfrac{5}{8}$ ◯ 0.675   <

*5. Write $2\dfrac{1}{4}$ as a percent.
225%

6. Write $4\dfrac{2}{5}$ as a percent.
440%

7. Write 0.7 as a percent.
70%

8. Write $\dfrac{7}{8}$ as a percent.
$87\frac{1}{2}$%

9. Twenty-five percent is the same as what fraction? $\dfrac{1}{4}$

10. $2^4 = $ 16

11. How much is $\dfrac{1}{8}$ of 360?
45

12. $6\dfrac{3}{4} + 5\dfrac{7}{8} = 12\dfrac{5}{8}$

13. $6\dfrac{1}{3} - 2\dfrac{1}{2} = 3\dfrac{5}{6}$

14. $2\dfrac{1}{2} \div 3 = \dfrac{5}{6}$

15. $6.93 + 8.429 + 12 = $ 27.359

16. $(1 - 0.1) \times (1 \div 0.1) = $ 9

17. $4.2 + \dfrac{7}{8} = $ (decimal answer)   5.075

18. $3\dfrac{1}{3} - 2.5 = $ (fraction answer) $\dfrac{5}{6}$

19. If 80% passed, then what percent did not pass?   20%

20. Compare: $\dfrac{1}{2} \div \dfrac{1}{3}$ ◯ $\dfrac{1}{3} \div \dfrac{1}{2}$ >

21. What is the next number in this sequence? 1000, 100, 10, 1, —— $\frac{1}{10}$ or 0.1

22. $3n = 4.2 - 1.5$ $n = 0.9$

23. The perimeter of the square is 48 inches. What is the area of one of the triangles?
72 sq. in.

24. How many years were there from 15 B.C. to 63 A.D.?
77 years

25. If Mark ran a 2-mile race, it would probably take him about how long? (a) $9\frac{1}{2}$ min. (b) 11 min. (c) 13 min. (d) 20 min.
(c)

MARK'S PERSONAL RUNNING RECORDS	
Distance	Time (Minutes:Seconds)
$\frac{1}{4}$ mile	0:58
$\frac{1}{2}$ mile	2:12
1 mile	6:05

LESSON 97

Fraction-Decimal-Percent Equivalents

Recall that there are three forms of fractions: common fractions, decimal fractions, and percents. We should be able to change from one form to another. In the following problem sets you will be asked to complete tables which show equivalent fractions, decimal fractions, and percents. As you complete these tables be sure to reduce fractions and simplify decimals.

example 97.1 Complete the table.

solution For the fraction $\frac{1}{2}$ we write a decimal and a percent. For the decimal 0.3 we write a fraction and a percent. For the percent 40% we write a fraction and a decimal.

	FRACTION	DECIMAL	PERCENT
1.	$\frac{1}{2}$	a.	b.
2.	a.	0.3	b.
3.	a.	b.	40%

1. a. $\dfrac{1}{2} = 2\overline{)1.0}^{\,0.5}$ **b.** $\dfrac{1}{2} \times \dfrac{100\%}{1} = \dfrac{100\%}{2} = 50\%$

2. a. $0.3 = \dfrac{3}{10}$ **b.** $0.3 \times 100\% = 30\%$

3. a. $40\% = \dfrac{40}{100} = \dfrac{2}{5}$ **b.** $40\% = 0.40 = 0.4$

practice
(See Practice Set II in the Appendix.)

Complete the table below.

FRACTION	DECIMAL	PERCENT
$\frac{3}{5}$	**a.** 0.6	**b.** 60%
c. $\frac{4}{5}$	0.8	**d.** 80%
e. $\frac{1}{5}$	**f.** 0.2	20%
$\frac{3}{4}$	**g.** 0.75	**h.** 75%
i. $\frac{3}{25}$	0.12	**j.** 12%
k. $\frac{1}{20}$	**l.** 0.05	5%

problem set 97

1. When the sum of $\frac{1}{2}$ and $\frac{1}{4}$ is divided by the product of $\frac{1}{2}$ and $\frac{1}{4}$, what is the quotient?   6

2. Jenny is 5.5 feet tall. She is how many inches tall?   66 in.

3. If 0.8 of the 200 runners finished the race, how many runners did not finish the race?   40

4. If BC is 36 cm and AC is 63 cm, then how long is AB?
   27 cm

5. What percent is equivalent to $1\frac{3}{10}$? 130%

6. Here is part of a multiplication-facts table, but one number is wrong. Which is the wrong number? 81

48	54
56	63
64	81

7. Round twenty-seven million, forty thousand, eight hundred to the nearest thousand. 27,041,000

8. What is the next number in this sequence? 1, 1, 2, 3, 5, 8, 13, — 21

9. If there is a 20% chance of rain, then what is the chance that it will not rain? 80%

10. $5^2 - 5 =$ 20

11. $(3,732 + 348 + 2,064) \div 12 =$ 512

12. $\dfrac{1 - 0.001}{0.03} =$ 33.3   **13.** $\dfrac{1\frac{1}{2}}{\frac{2}{3}} =$ $2\frac{1}{4}$

14. $0.36 \times 0.47 \times 100 =$ 16.92

15. $6\frac{1}{2} + 4.95 =$ (decimal answer) 11.45

16. $12\frac{1}{6} - 3.5 =$ (fraction answer) $8\frac{2}{3}$

17. Round 48.3757 to the nearest hundredth. 48.38

18. What fraction of a foot is 3 inches? $\frac{1}{4}$

19. What decimal part of a meter is 3 centimeters? 0.03

20. What is the place value of the 3 in 123,456.789?
thousands

21. Arrange in order from least to greatest: $-1, \frac{1}{2}, -\frac{1}{2}$
$-1, -\frac{1}{2}, \frac{1}{2}$

22. These two triangles together form what shape?   trapezoid

Complete the chart to answer problems 23, 24, and 25.

	FRACTION	DECIMAL	PERCENT
23.	$\frac{1}{2}$	**a.** 0.5	**b.** 50%
24.	**a.** $\frac{3}{10}$	0.3	**b.** 30%
25.	**a.** $\frac{2}{5}$	**b.** 0.4	40%

LESSON 98

Volume

The **volume** of a shape is how much space it "takes up." To measure **volume** we must use units which take up space. The units we use are cubes:

like cubic centimeters or cubic inches. The volume of a shape is the number of cubic units of space it occupies. We will use sugar cubes to help us keep this idea in mind.

How many sugar cubes are needed to form this shape?

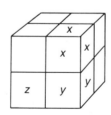

We must be careful when counting. We want to count the total number of single cubes. We can see three faces of the x cube, two faces of the y cube, and one face of the z cube. Are there any cubes which we cannot see? One way to approach the problem is to separate the cube into layers and count the number of single cubes in each layer. In the picture on the preceding page, we see two layers and four cubes on the top layer. Two layers of four cubes makes a total of **eight** sugar cubes in the shape.

practice
a. How many cubes are on the top layer? 9
b. How many layers are there in all? 3
c. How many single cubes are there in all? 27
d. If each cube is one cubic centimeter, then what is the volume of this shape? 27 cu. cm

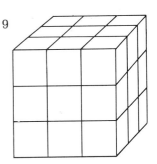

problem set 98

1. A foot-long hot dog can be cut into how many $1\frac{1}{2}$-inch lengths? 8

2. A can of beans is the shape of what geometric solid? (Name the shape.) cylinder

3. If $\frac{3}{8}$ of them voted yes and $\frac{3}{8}$ voted no, then what fraction did not vote? $\frac{1}{4}$

4. Nine months is (a) what fraction of a year? (b) what decimal part of a year? (a) $\frac{3}{4}$ (b) 0.75

5. This rectangular prism is made up of how many cubes? 12

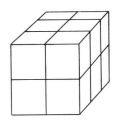

6. What percent is equal to 2.5? 250%

7. If $\frac{1}{5}$ of the pie was eaten, what percent of the pie was left?
80%

8. Write the percent form of $\dfrac{1}{7}$. $14\frac{2}{7}\%$

9. $3^2 - 2^3 =$ 1   **10.** $5 \cdot 4 \cdot 3 \cdot 2 \cdot 1 \cdot 0 =$ 0

11. $\dfrac{4.5}{0.18} =$ 25

12. $6\dfrac{2}{3} + 6.2 =$ (fraction answer) $12\dfrac{13}{15}$

13. $6\dfrac{3}{4} - 6.2 =$ (decimal answer)   0.55

14. $12\dfrac{1}{2} \times 1\dfrac{3}{5} \times 5 =$ 100   **15.** $(4.2 \times 0.05) \div 14 =$ 0.015

16. $\dfrac{1\frac{1}{2}}{3} = \dfrac{1}{2}$

17. Round $7.7777 to the nearest cent. $7.78

18. Compare: $1\dfrac{7}{8} + 2\dfrac{5}{6} + 3\dfrac{4}{5} \bigcirc 2 + 3 + 4$ <

19. Write the prime factorization of 90. $2 \cdot 3 \cdot 3 \cdot 5$

20. $10n = 1.2$ $n =$ 0.12

21. What is the greatest common factor of 40 and 60?   20

22. The perimeter of a square is 2 meters. How many centimeters long is each side?   50 cm

23. $\dfrac{1}{4} - \dfrac{1}{2} = -\dfrac{1}{4}$

Complete the chart to answer problems 24 and 25.

	FRACTION	DECIMAL	PERCENT
24.	**a.** $\frac{4}{5}$	0.8	**b.** 80%
25.	**a.** $\frac{1}{20}$	**b.** 0.05	5%

LESSON
99

Volume of a Rectangular Prism

Here is a picture of a solid which has faces that are rectangles. This shape is known as a **rectangular prism**. (Sometimes it is called a **rectangular solid** or **right prism**.)

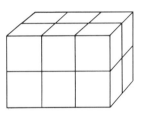

Solids take up space. The amount of space a solid occupies is called its **volume**. To measure volume we must use units which occupy space. The units we use are cubes. We will use sugar cubes of given sizes to help us keep this idea in mind.

This rectangular prism is formed by 2 layers of sugar cubes with 6 cubes in each layer. Since 12 sugar cubes were used we say that its volume is 12 **cubic** units.

example 99.1 How many sugar cubes 1 inch on an edge would be needed to form this rectangular prism?

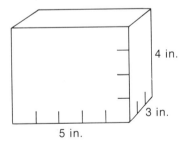

4 in.

3 in.

5 in.

solution The area of the base is 3 inches times 5 inches for 15 square inches. Thus we can set 15 sugar cubes on the bottom layer as we show. The solid is 4 inches high, so we will have 4 layers for a total of 4 times 15, or **60 sugar cubes** in all.

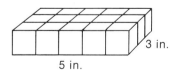

3 in.

5 in.

example 99.2 What is the volume of a cube whose edges are 10 centimeters long?

solution The area of the base is 10 cm times 10 cm, or 100 sq. cm. Thus we can set 100 sugar cubes on the bottom layer. There are 10 layers so there are 10 times 100, or 1000 sugar cubes in all. Thus the volume is **1000 cu. cm.**

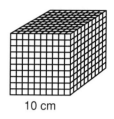

10 cm

problem set 99

1. If 0.6 is the divisor and 1.2 is the quotient, then what is the dividend?   0.72

2. If a number is twelve less than fifty, then it is how much more than twenty?   18

3. If the sum of 4 numbers is 14.8, then the average of the 4 numbers is what?   3.7

*4. How many sugar cubes 1 inch on an edge would be needed to form this rectangular prism?   60

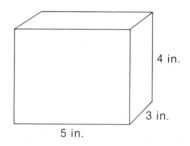

4 in.

3 in.

5 in.

*5. What is the volume of a cube which has edges 10 cm long?
   1000 cu. cm

6. How many cubic-inch blocks are needed to fill a rectangular box which is 10 inches long, 5 inches wide, and 4 inches high?   200

7. Write the numeral nineteen million, two hundred five thousand, sixty and three hundredths.   19,205,060.03

8. What is the percent equivalent of 2.1?   210%

9. The fraction $\frac{2}{3}$ is equal to what percent? $66\frac{2}{3}\%$

10. If 20% of the students earned A's, then what fraction of the students did not earn A's? $\frac{4}{5}$

11. $4^2 - 2^4 =$ 0   **12.** $5\frac{7}{8} + 4\frac{3}{4} = 10\frac{5}{8}$ **13.** $\dfrac{1\frac{1}{2}}{2\frac{1}{2}} = \dfrac{3}{5}$

14. $8.47 + 9 + 4.6 =$ 22.07   **15.** $8.75 \times 1.6 =$ 14

16. $7.2 \div 1.5 =$ 4.8   **17.** $198.5 \div 100 =$ 1.985

18. $3.18 - 3\frac{1}{8} =$ (decimal answer)   0.055

19. Arrange in order from **greatest** to **least**: 1.2, $\dfrac{3}{2}$, 100%
$\frac{3}{2}$, 1.2, 100%

20. What is the least common multiple of 9 and 12?   36

21. Round 0.16225 to the nearest thousandth.   0.162

22. $8.4 + x = 12$ $x =$ 3.6

23. The perimeter of the triangle is 24 cm. What is the area of the triangle?
24 sq. cm

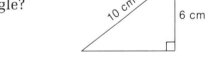

24. $575 - 1000 =$ −425

25. How long is the line segment?
$1\frac{5}{8}$ in.

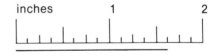

Radius, Diameter, Circumference

Some plane shapes are formed with straight lines. Other plane shapes are formed with curved lines. One shape formed with a curved line is the **circle**. When we consider a circle, we may refer to the distance around the circle, the distance across the circle, and the distance to the circle from its center.

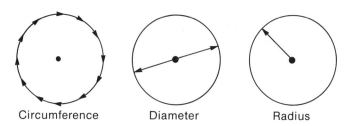

Circumference   Diameter   Radius

The **circumference** is the distance **around** the circle. This distance is the same as the perimeter of a circle. The circumference always means the distance around a circle. The perimeter is the distance around a circle or around any figure.

The **diameter** is the distance **across** a circle through the center.

The **radius** is the distance from the center to the circle. The plural of radius is **radii**. A diameter is the same as two radii.

example 100.1   What is the name for the perimeter of a circle?

solution   The distance around a circle is its **circumference**.

example 100.2   If the radius is 4 cm, what is the length of the diameter?

solution   The diameter of a circle is twice its radius—in this case, **8 cm**.

practice   Give the name for:
a. The distance across a circle   diameter
b. The distance around a circle   circumference
c. The distance from the center to the circle   radius
d. If the diameter is 10 in., what is the radius?   5 in.

problem set 100

1. If the cost of calling Miami from Denver is $1.48 for the first 3 minutes plus 35¢ for each additional minute, then what would be the total cost of a 10-minute call? $3.93

2. A shoe box is the shape of what geometric solid?
rectangular prism

3. If the average of six numbers is 4.2, then what is the sum of the six numbers?   25.2

4. If AB is $\frac{1}{4}$ of AC, and if AC is 12 cm, then how long is BC?
9 cm

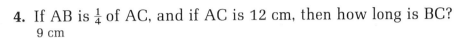

*5. What is the name for the perimeter of a circle?
circumference

*6. If the radius is 4 cm, what is the length of the diameter?
8 cm

7. A diameter is equal to how many radii?   2

8. A rectangular solid which is 5 cm long, 4 cm wide, and 3 cm high has a volume of how many cubic centimeters?
60 cu. cm

9. How many is 50% of one dozen?   6

Complete the chart to answer problems 10, 11, and 12.

	FRACTION	DECIMAL	PERCENT
10.	$\frac{3}{4}$	**a.** 0.75	**b.** 75%
11.	**a.** $\frac{9}{10}$	0.9	**b.** 90%
12.	**a.** $\frac{1}{25}$	**b.** 0.04	4%

13. $3\frac{5}{8} + 2\frac{1}{2} = 6\frac{1}{8}$ **14.** $90 - 64\frac{1}{4} = 25\frac{3}{4}$

15. $2 − $1.42 = ¢   58¢   **16.** 16¢ × 14 = $ $2.24

17. $5\frac{1}{2} \times 4 = 22$ **18.** $5\frac{1}{2} \div 4 = 1\frac{3}{8}$

19. $0.12 \div 8 = 0.015$ **20.** $9 \div 0.12 = 75$

21. Compare: $6.142 \times 9.065 \bigcirc 54$ $>$

22. $6n = 23.4$ $n = 3.9$

23. These three triangles together form what shape?   pentagon

24. At 6 A.M. the temperature was $-8°$ F. By noon the temperature was $15°$ F. The temperature had risen how many degrees? $23°$ F

25. To what decimal number is the arrow pointing?   8.8

LESSON 101

Comparing Fractions— Cross Products

Fractions can be compared easily when they have common denominators, for if the denominators are the same, then it is only necessary to compare the numerators. One way to rewrite two fractions so that they will have common denominators is to multiply each fraction by a name for one which uses the denominator of the other fraction. We can rewrite $\frac{3}{5}$ and $\frac{4}{7}$ with common denominators by multiplying $\frac{3}{5}$ by $\frac{7}{7}$ and $\frac{4}{7}$ by $\frac{5}{5}$.

$$\frac{3}{5} \times \frac{7}{7} = \frac{21}{35} \qquad \frac{4}{7} \times \frac{5}{5} = \frac{20}{35}$$

Since $\frac{21}{35}$ is greater than $\frac{20}{35}$, the same must be true of their equivalent fractions: $\frac{3}{5} > \frac{4}{7}$.

We will write this another way to illustrate a point.

$$\frac{\textcircled{21}}{35}, \frac{3}{5} > \frac{4}{7}, \frac{\textcircled{20}}{35}$$

Notice that the numerator 21 was formed by multiplying 3 by 7, and that the numerator 20 was formed by multiplying 4 by 5. We can abbreviate this work as shown below.

$$\textcircled{21} \ \frac{3}{5} \times \frac{4}{7} \ \textcircled{20}$$

These numbers, 21 and 20, are sometimes called **cross products** because they can be found by criss-cross multiplication. Since these cross products are actually the numerators of fractions which have common denominators, we may simply compare the cross products of two fractions to compare the fractions themselves. If the cross products are equal, the fractions are equal; if the cross products are not equal, then the fraction with the greater cross product is the greater fraction.

example 101.1 Compare: $\frac{3}{8} \bigcirc \frac{2}{5}$

solution Using cross products is a quick way to compare fractions. Comparing these fractions, we see that the cross product at $\frac{2}{5}$ is greater than the cross product at $\frac{3}{8}$, so

$$\textcircled{15} \ \frac{3}{8} \times \frac{2}{5} \ \textcircled{16}$$

$$\frac{3}{8} < \frac{2}{5}$$

problem set 101

1. What is the reciprocal of $3\frac{1}{3}$? $\frac{3}{10}$

2. On his first four tests Chris had scores of 92%, 96%, 92%, and 84%. What was his test average after four tests? 91%

3. In basketball, one point is scored for a free throw and two points for a field goal. If a team scored 96 points while making 18 free throws, how many field goals were made? 39

4. Compare: $\frac{3}{5} \bigcirc \frac{9}{15}$ =

*5. Compare: $\frac{3}{8} \bigcirc \frac{2}{5}$ <

6. Arrange in order of size from least to greatest: $-1, 1, 0.1$
$-1, 0.1, 1$

7. $10^3 - 10^2 = 900$

8. What fraction of the diameter of a circle is its radius?
$\frac{1}{2}$

Complete the chart to answer problems 9, 10, and 11.

	FRACTION	DECIMAL	PERCENT
9.	$\frac{4}{25}$	**a.** 0.16	**b.** 16%
10.	**a.** $\frac{1}{100}$	0.01	**b.** 1%
11.	**a.** $\frac{9}{10}$	**b.** 0.9	90%

12. $1\frac{2}{3} + 3\frac{1}{2} + 4\frac{1}{6} = 9\frac{1}{3}$

13. $\frac{5}{6} \times \frac{3}{10} \times 4 = 1$

14. $3\frac{1}{2} \div 2 = 1\frac{3}{4}$

15. $6.437 + 12.8 + 7 = 26.237$

16. $4.3 \times 0.0067 = 0.02881$

17. Divide 9.4 by 8 and round the quotient to the nearest tenth. 1.2

18. An octagon has how many more sides than a pentagon?
3

19. $20 + 25 + n = 60$ $n = 15$

20. $2 \cdot m = 3 \cdot 6$ $m = 9$

21. How many cubes 1 inch on each edge would be needed to build this larger cube? 64 cu. in.

4 in.

22. What is 0.3 of 60? 18

23. The average of four numbers is 5. What is their sum?
20

24. How many years were there from 5 B.C. to 28 A.D.?
32 years

25. How many millimeters long is the line segment?   35 mm

LESSON
102

Ratio

A **ratio** is a way of showing a relationship between two or more numbers. If there were 13 boys and 15 girls in a classroom, then the boy-girl ratio would be 13 to 15. If a team won 7 games and lost 3, then its won-lost ratio would be 7 to 3.

Ratios can be written many ways. They can be written as fractions or as decimal numbers. They can be written by using colons. They can also be written in word form. We will concentrate on the fraction form of a ratio. A ratio of 13 to 15 is written in fraction form as $\frac{13}{15}$. A ratio of 7 to 3 is written $\frac{7}{3}$. Notice that we do not convert ratios to mixed numbers. We usually write ratios in reduced form, just as we usually write fractions in reduced form.

example 102.1   Write the fraction form of the ratio 4 to 6.

solution   The ratio 4 to 6 is written $\frac{4}{6}$, which should be reduced to $\frac{2}{3}$.

example 102.2   In a class of 30 children there are 13 boys. What is the boy-girl ratio?

solution   To name the boy-girl ratio we must first find the number of girls. If 13 of the 30 children are boys, then there must be 17 girls. The boy-girl ratio is $\frac{13}{17}$. (Notice that ratios must be named **in the order given**. The girl-boy ratio in this case would be $\frac{17}{13}$.)

practice
(See Practice Set JJ in the Appendix.)

Write the fraction form of these ratios. Reduce when possible.

a. 6 to 10 $\frac{3}{5}$

b. 10 to 6 $\frac{5}{3}$

c. 8 to 12 $\frac{2}{3}$

d. 20 to 15 $\frac{4}{3}$

e. What is the dog-cat ratio in a neighborhood which has 18 cats and 12 dogs? $\frac{2}{3}$

f. What is the girl-boy ratio in a class of 29 students which has 15 boys? $\frac{14}{15}$

problem set 102

1. When the sum of $\frac{1}{2}$ and $\frac{1}{3}$ is divided by the product of $\frac{1}{2}$ and $\frac{1}{3}$, what is the quotient?   5

2. The average age of three men is 24. If two of the men are 22 years old, how old is the third?   28

3. A string one yard long is formed into the shape of a square. How many square inches is the area of the square?
   81 sq. in.

*4. Write the fraction form of the ratio 4 to 6. $\frac{2}{3}$

*5. In a class of 30 students there are 13 boys. What is the boy-girl ratio of the class? $\frac{13}{17}$

6. $100 - 10^2 = 0$

7. The diameter of a circle is 2 feet. The radius is how many inches?   12 in.

8. Compare: $\dfrac{5}{8}\ \bigcirc\ \dfrac{7}{11}$ <

Complete the chart to answer problems 9, 10, and 11.

	FRACTION	DECIMAL	PERCENT
9.	$\frac{1}{100}$	**a.** 0.01	**b.** 1%
10.	**a.** $\frac{2}{5}$	0.4	**b.** 40%
11.	**a.** $\frac{2}{25}$	**b.** 0.08	8%

12. $7\dfrac{1}{2} + 6\dfrac{3}{4} + 1\dfrac{1}{8} = 15\dfrac{3}{8}$ **13.** $7\dfrac{1}{2} - 1\dfrac{3}{4} = 5\dfrac{3}{4}$

14. $2 \div 3\dfrac{1}{2} = \dfrac{4}{7}$ **15.** $(6 + 2.4) \div 0.04 = 210$

16. $12.4 \times 6.2 \times 100 = 7688$

17. Divide 3.91 by 4 and round the quotient to the nearest hundredth.   0.98

18. Write the numeral twenty million, five hundred seven thousand, ninety-six.   20,507,096

19. List the prime numbers between 40 and 50.   41, 43, 47

20. $4 \cdot 6 = 3 \cdot n$ $n = 8$

21. The perimeter of the triangle is 60 mm. What is its area?   150 sq. mm

22. Seven nickels is what decimal part of a dollar?   0.35

23. Seven nickels is what fraction of a dollar? $\dfrac{7}{20}$

24. How many centimeters is 2.2 meters?   220 cm

25. What temperature is shown on the thermometer? $-8°$ F

LESSON
103

Proportions

proportions A **proportion** is a true statement that two ratios are equal. An example of a proportion is $\frac{3}{4} = \frac{6}{8}$.

example 103.1 Which ratio forms a proportion with $\frac{2}{3}$? $\frac{2}{4}, \frac{3}{4}, \frac{4}{6}, \frac{3}{2}$

solution Equal ratios form a proportion. The ratio equal to $\frac{2}{3}$ is **$\frac{4}{6}$**.

completing proportions Most proportion problems involve finding a missing number in a proportion. The missing number is often shown by a letter, such as $\frac{3}{5} = \frac{6}{a}$. We can find the missing number in a proportion by using cross products.

example 103.2 Complete this proportion: $\frac{3}{5} = \frac{6}{a}$

solution The equals sign shows that the two ratios form a proportion. We must find the number which completes the second ratio. Since equal fractions have equal cross products, we may cross-multiply to form a new equation: $3a = 30$. This new equation may be solved by dividing by the known factor, in this case 3.

Step 1.   Cross-multiply. $\frac{3}{5} = \frac{6}{a}$ $3a = 30$

Step 2.   Divide by the known factor, in this case 3. $a = \frac{30}{3} = \mathbf{10}$

example 103.3 What is the missing term in this proportion? $\frac{6}{8} = \frac{w}{12}$

solution We will solve this proportion by using cross products.

Step 1.   Cross-multiply. $\dfrac{6}{8} = \dfrac{w}{12}$ $72 = 8w$

Step 2.   Divide by the known factor, which is 8. $w = \dfrac{72}{8} = \mathbf{9}$

practice
(See Practice
Set KK in the
Appendix.)

Find the missing term in each proportion.

a. $\dfrac{3}{5} = \dfrac{c}{25}$ 15

b. $\dfrac{5}{f} = \dfrac{15}{21}$ 7

c. $\dfrac{9}{12} = \dfrac{6}{w}$ 8

d. $\dfrac{6}{4} = \dfrac{r}{12}$ 18

e. $\dfrac{m}{15} = \dfrac{6}{10}$ 9

f. $\dfrac{4}{y} = \dfrac{12}{24}$ 8

**problem
set 103**

1. A pyramid with a square base has how many more edges than vertexes?   3

2. Mark had nickels, dimes, and quarters in his pocket. He had half as many dimes as nickels and half as many quarters as dimes. If he had 4 dimes, then how much money did he have in his pocket? $1.30

3. What number is five more than the product of six and seven?   47

4. A team won 6 games and lost 10. What was its won-lost ratio? $\frac{3}{5}$

*5. Which ratio forms a proportion with $\dfrac{2}{3}$? $\dfrac{2}{4}, \dfrac{3}{4}, \dfrac{4}{6}, \dfrac{3}{2}$
$\frac{4}{6}$

*6. Solve for the missing term: $\dfrac{6}{8} = \dfrac{a}{12}$ 9

7. What is the perimeter of the hexagon? (All units are centimeters.)
50 cm

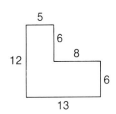

8. Compare: $\dfrac{6}{9} \bigcirc \dfrac{10}{12}$ <

Complete the chart to answer questions 9, 10, and 11.

	FRACTION	DECIMAL	PERCENT
9.	$\frac{3}{20}$	**a.** 0.15	**b.** 15%
10.	**a.** $1\frac{1}{5}$	1.2	**b.** 120%
11.	**a.** $\frac{1}{10}$	**b.** 0.1	10%

12. $\left(3\frac{1}{10} + 1\frac{1}{5}\right) - 2\frac{1}{2} = 1\frac{4}{5}$ **13.** $6\frac{2}{3} \times 2\frac{1}{10} = 14$

14. $1\frac{2}{3} \div 1\frac{1}{2} = 1\frac{1}{9}$ **15.** $(1 - 0.2) - 0.03 = 0.77$

16. $0.5 \times 0.5 \times 0.5 \times 0.5 = 0.0625$

17. Divide 0.624 by 0.05 and round the quotient to the nearest whole number.   12

18. The average of three numbers is 20. What is the sum of the three numbers?   60

19. Write the prime factorization of 45. $3 \times 3 \times 5$

20. $5 + n + 5 = 18$ $n = 8$ **21.** $3^4 + y = 81$ $y = 0$

22. How many blocks 1 inch on a side would it take to fill a shoe box which is 12 inches long, 6 inches wide, and 5 inches high?   360

23. Three fourths of the 60 players played. How many did not play?   15

24. Tom lives 15 kilometers south of Jim. Jerry lives 20 kilometers north of Jim. How many kilometers from Jerry does Tom live?   35 km

25. The distance a car travels can be found by multiplying the **speed** of the car times the amount of **time** the car travels at that speed. How far would a car travel in 4 hours at 88 kilometers per hour?   352 km

LESSON
104

Ratio Word Problems

Proportions can be used to solve many types of problems. In this book we will use proportions to solve ratio word problems.

example 104.1   The ratio of salamanders to frogs was 5 to 7. If there were 20 salamanders, how many frogs were there?

solution   In this problem there are two kinds of numbers. There are ratio numbers, and there is an actual count of salamanders. We arrange these numbers in a table or "ratio box." We use the letter f to stand for the actual count of frogs.

	RATIO	ACTUAL COUNT
Salamanders	5	20
Frogs	7	f

From this table we write a proportion. Then we solve the proportion and find that there are **28 frogs**.

	RATIO	ACTUAL COUNT
Salamanders	5	20
Frogs	7	f

$$\longrightarrow \quad \frac{5}{7} = \frac{20}{f}$$

$$5f = 140, \text{ so } f = \mathbf{28}$$

example 104.2   The ratio of humpback whales to killer whales was 2 to 7. If there were 42 killer whales, how many humpbacks were there?

solution   We make a ratio box. Then we write a proportion.

	RATIO	ACTUAL COUNT
Humpback	2	h
Killer	7	42

$$\longrightarrow \quad \frac{2}{7} = \frac{h}{42}$$

We will solve the proportion by using cross products.

$$7h = 84 \qquad \text{so } h = \textbf{12}$$

practice   Solve these ratio problems. Begin by making a ratio box.
  a. There were more dragons than knights in the battle. In fact, the ratio of dragons to knights was 5 to 4. If there were 60 knights, how many dragons were there?   75
  b. At the party the boy-girl ratio was 5 to 3. If there were 30 boys, how many girls were there?   18

**problem
set 104**

  1. How far would a car travel in $2\frac{1}{2}$ hours at 50 miles per hour?   125 mi.

  2. A map is drawn to this scale: 1 inch = 2 miles. How many miles apart are two towns which are 3 inches apart on the map?   6 mi.

  *3. The ratio of humpback whales to killer whales was 2 to 7. If there were 42 killer whales, how many humpback whales were there? Begin by making a ratio box.   12

  4. One way to make 10¢ is with 10 pennies. Another way is with 2 nickels. Altogether, **how many** ways are there to make 10¢?   4

  5. In the class of 30 students there were 12 boys. What was the boy-girl ratio? $\frac{2}{3}$

  6. Which of the following is not a quadrilateral? parallelogram, pentagon, rhombus   pentagon

  7. Marge swung her arm in a circle. If her arm is 24 inches long, then what is the diameter of the circle?   48 in.

  8. Arrange in order from least to greatest: $\dfrac{1}{3}, \dfrac{2}{5}, \dfrac{3}{10}$ $\frac{3}{10}, \frac{1}{3}, \frac{2}{5}$

Complete the chart to answer problems 9, 10, and 11.

	FRACTION	DECIMAL	PERCENT
9.	$\frac{3}{50}$	**a.** 0.06	**b.** 6%
10.	**a.** $\frac{1}{25}$	.04	**b.** 4%
11.	**a.** $1\frac{1}{2}$	**b.** 1.5	150%

12. $4\dfrac{1}{12} + 5\dfrac{1}{6} + 2\dfrac{1}{4} = \quad 11\dfrac{1}{2}$ **13.** $\dfrac{4}{5} \times 3\dfrac{1}{3} \times 3 = \quad 8$

14. Solve: $\dfrac{c}{12} = \dfrac{3}{4}$ 9

15. $(1 + 0.2) \div (1 - 0.2) = \quad 1.5$ **16.** $0.125 \times 80 = \quad 10$

17. Round the quotient to the nearest hundredth: $0.875 \div 4 = $ 0.22

18. Write the decimal numeral one hundred five and five-hundredths.   105.05

19. List the factors of 50.   **20.** $25n = 25^2$ $n = 25$
1, 2, 5, 10, 25, 50

21. A nickel is what decimal part of a dime?   0.5

22. What is $\dfrac{1}{4}$ of 360?   90

23. The perimeter of the triangle is 18 cm. What is its area?   12 sq. cm

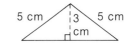

24. The temperature was $-5°$ F at 6:00 A.M. By noon the temperature had risen 12 degrees. What was the noontime temperature?   7° F

25. How many eighths equal three fourths?   6

LESSON 105

Measuring to One Sixteenth of an Inch

The inch ruler may be divided into smaller and smaller sections. The inches are divided into halves, then those halves are divided into fourths (quarters), and so on.

Dividing a whole into 2 parts makes halves.

Dividing a half into 2 parts makes quarters.

Dividing a quarter into 2 parts makes eighths.

Dividing an eighth into 2 parts makes sixteenths.

We have measured to eighths of an inch. With this lesson we begin measuring to sixteenths of an inch.

Each inch has been divided into 16 equal lengths. Each of these lengths is $\frac{1}{16}$ of an inch. To measure to sixteenths of an inch we may count the number of small lengths and reduce when necessary. With practice we should be able to identify the measure without needing to count each small mark.

practice   To which mark is each arrow pointing?

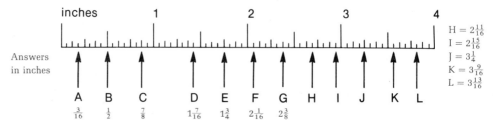

$H = 2\frac{11}{16}$
$I = 2\frac{15}{16}$
$J = 3\frac{1}{4}$
$K = 3\frac{9}{16}$
$L = 3\frac{13}{16}$

m. How many sixteenths equal a fourth?   4

n. Three eighths is the same as how many sixteenths?   6

o. Eight sixteenths is the same as how many halves?   1

problem set 105

1. The average of three numbers is 20. If the greatest is 28, and the least is 15, then what is the third? 17

2. On a map drawn to the scale of 1 inch = 10 miles, how far apart are two points which on the map are $2\frac{1}{2}$ inches apart? 25 mi.

3. What fraction of a quarter is a dime? $\frac{2}{5}$

4. What decimal part of a quarter is a nickel? 0.2

5. $x^2 = 36$ $x = 6$

6. How many sixteenths equal one-half? 8

7. The circumference of each tire on John's bike is 6 feet. How many turns does each wheel make as John rides down his 30-foot-long driveway? 5

8. The ratio of kangaroos to koalas was 9 to 5. If there were 414 kangaroos, how many koalas were there? 230

Complete the chart to answer questions 9, 10, and 11.

	FRACTION	DECIMAL	PERCENT
9.	$\frac{1}{8}$	**a.** 0.125	**b.** $12\frac{1}{2}\%$
10.	**a.** $1\frac{4}{5}$	1.8	**b.** 180%
11.	**a.** $\frac{3}{100}$	**b.** 0.03	3%

12. $8\frac{1}{3} - 3\frac{1}{2} =$ $4\frac{5}{6}$

13. $2\frac{1}{2} \times 1\frac{1}{3} \times 1\frac{1}{5} =$ 4

14. $2\frac{1}{2} \div 1\frac{1}{4} = 2$

15. $1 - (0.2 \times 0.3) =$ 0.94

16. $0.014 \div 0.5 =$ 0.028

17. Round the product to the nearest thousandth. $1.42 \times 0.26 =$ 0.369

18. Write the standard notation for $(6 \times 10{,}000) + (9 \times 100) + (7 \times 1)$. 60,907

19. The number 100 has how many factors?   9

20. A 1-foot ruler broke into two pieces so that one piece was $5\frac{1}{4}$ inches long. How long was the other piece? $6\frac{3}{4}$ in.

21. $(14 \times 16) - (15 \times 15) =$ -1

22. How many blocks are in this shape?
24

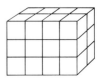

23. What percent of a meter is 20 centimeters?   20%

24. What time is 90 minutes after 11:15 A.M.?   12:45 P.M.

25. How long is the line segment?
$1\frac{9}{16}$ in.

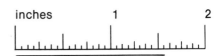

LESSON
106

Perimeter and Area of Complex Shapes

In earlier lessons we practiced finding the perimeter and area of simple shapes like rectangles and triangles. In this lesson we will practice finding the perimeter and area of complex shapes.

The figure below is an example of a complex shape. We can find the perimeter and area of a complex shape by using reason and imagination. Try to figure out the perimeter and area of this shape.

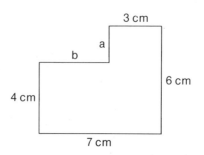

To find the perimeter of this shape we must first find the length of sides a and b. We see that the total height of the figure is 6 cm, but 4 cm plus a is also the height, so side a must equal 2 cm. Since the width is 7 cm, and side b plus 3 cm is equal to the width, side b must equal 4 cm. Adding all sides we find the perimeter is 26 cm.

We must use some imagination to find the area of the shape. One way to find the area is to imagine that the shape is cut into two rectangles. We can find the area of each rectangle, then add to find the total area of the shape, which is 34 sq. cm.

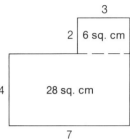

practice   Find the perimeter and area of each shape.

(a) $P = 32$ in.; $A = 40$ sq. in.
(b) $P = 24$ ft.; $A = 30$ sq. ft.
(c) $P = 70$ mm; $A = 268$ sq. mm

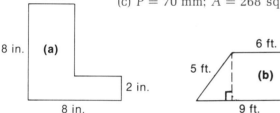

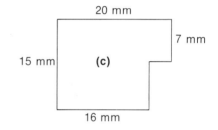

problem set 106

1. If the divisor is eight tenths and the dividend is twenty and eight hundredths, then what is the quotient?   25.1

2. The plans for the clubhouse were drawn so that 1 inch equaled 2 feet. On the plans the clubhouse was 4 inches tall. The actual clubhouse will be how tall?   8 ft.

3. If all the king's horses total 600, and all the king's men total 800, then what is the ratio of men to horses? $\frac{4}{3}$

4. What decimal part of the perimeter of a square is the length of one side?   0.25

5. What is the area of a room which is 12 feet long and 11 feet wide?   132 sq. ft.

6. $10^3 - (10^2 - 10) = $ 910

7. $(6 \times 100) + (4 \times 10) + (2 \times \frac{1}{10}) = $ 640.2

8. What number completes the proportion? $\dfrac{6}{n} = \dfrac{8}{12}$ 9

Complete the chart to answer problems 9, 10, and 11.

	FRACTION	DECIMAL	PERCENT
9.	$1\frac{1}{10}$	**a.** 1.1	**b.** 110%
10.	**a.** $\frac{9}{20}$	0.45	**b.** 45%
11.	**a.** $\frac{4}{5}$	**b.** 0.8	80%

12. $5\dfrac{3}{8} + 4\dfrac{1}{4} + 3\dfrac{1}{2} = 13\dfrac{1}{8}$ **13.** $\dfrac{8}{3} \times \dfrac{5}{12} \times \dfrac{9}{10} = 1$

14. $64.8 + 8.42 + 24 = $ 97.22   **15.** $6.25 \times 0.08 \times 100 = $ 50

16. How many sixteenths equal three fourths?   12

17. Round the quotient to the nearest hundredth: $0.625 \div 0.8 = $ 0.78

18. Write the number one hundred ten million, nine hundred eighty thousand.   110,980,000

19. What is the greatest common factor of 30 and 45?   15

20. Compare: $\dfrac{24}{25} \bigcirc \dfrac{25}{26}$ <

21. How many blocks with edges 1 inch long would be needed to fill a box with all edges 1 foot long?   1728

22. $6 - a = -4$ $a = 10$

23. What is the perimeter of this shape?   30 cm

24. What is the area of this shape?
41 sq. cm

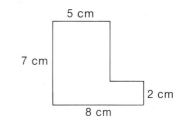

5 cm

7 cm

2 cm

8 cm

25. How long is the line segment?
$\frac{15}{16}$ in.

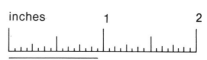

inches   1   2

LESSON
107

Acute Angles, Right Angles, Obtuse Angles, Straight Angles

The words **acute**, **right**, **obtuse**, and **straight** are terms we use to name angles of various sizes.

A square corner is a **right angle**. Doors, chalkboards, windows, books, papers, and buildings all contain right angles.

An angle smaller than a right angle is an **acute angle**. Some remember this by thinking of "a cute little angle."

An angle greater than a right angle is an **obtuse angle**. An angle which makes a straight line is a **straight angle**. The angles drawn below illustrate each type.

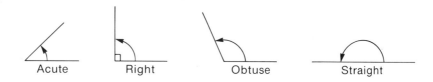

Acute   Right   Obtuse   Straight

A unit we use for measuring angles is a degree. A full circle or a turn "all the way around" measures 360 degrees. We abbreviate the word **degree** with a small circle above and to the right of the number, as in 360°. A right angle is $\frac{1}{4}$ of a full circle. One fourth of 360° is 90°. A right angle measures 90°. An acute angle measures less than 90°. An obtuse angle measures more than 90° but less than 180°. A straight angle is a half turn and measures 180°.

example 107.1   If angle x is an acute angle, then what type of angle is angle y?

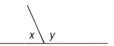

solution   We see that angle x is smaller than a right angle and that angle y is greater than a right angle. Angle y is an **obtuse** angle.

problem set 107

1. What two-digit number is a multiple of 5 and a factor of 125? 25

2. A set of house plans was drawn to this scale: 1 in. = 2 ft. On the plans a room measures 5 inches by 6 inches. What is the area of the room? 120 sq. ft.

3. How many **minutes** are there from 10:30 in the morning to a quarter after one in the afternoon? 165 min.

4. $3^4 - 4^3 =$ 17

5. Reduce: $\dfrac{36}{54}$ $\dfrac{2}{3}$

6. Write the prime factorization of 84. 2 × 2 × 3 × 7

7. Ol' McDonald's horse was tied to a post in the center of the pasture. What is the shape of the region in which the horse may roam? circle

8. What number completes the proportion? $\dfrac{2}{3} = \dfrac{8}{n}$ 12

Complete the chart to answer problems 9, 10, and 11.

	FRACTION	DECIMAL	PERCENT
9.	$\frac{3}{8}$	**a.** 0.375	**b.** $37\frac{1}{2}\%$
10.	**a.** $\frac{3}{50}$	0.06	**b.** 6%
11.	**a.** $1\frac{2}{5}$	**b.** 1.4	140%

12. $5 - \left(6\frac{1}{4} - 2\frac{1}{2}\right) =$ $1\frac{1}{4}$

13. $3\frac{3}{4} \times 2\frac{2}{5} \times 1\frac{1}{3} =$ 12

14. $7 - (4.1 - 0.42) =$ 3.32

15. $0.195 \times 0.048 =$ 0.00936

16. $0.0195 \div 30 =$ 0.00065

17. $6n = 2.1$ $n =$ 0.35

18. Round the quotient to the nearest cent: $\$8.78 \div 5 =$
$1.76

19. What is the least common multiple of 9 and 12?   36

*20. If angle x is an acute angle, then what type of angle is angle y?   obtuse

21. What type of an angle is an angle which measures 89°?
acute

22. The perimeter of the triangle is 16 cm.
What is its area?   12 sq. cm

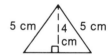

23. $3a + 1 = 25$ $a =$ 8

24. How many cubes make this shape?
48

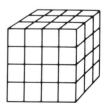

25. What number is halfway between 82 and 28 on the number line?   55

LESSON
108

"Similar"

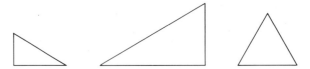

Two of these shapes are **similar**. In geometry the word **similar** means that two figures are the same shape but not necessarily the same size. When the similar figures are polygons, then all the angles of the two polygons will "match." That is, each of the angles of one polygon will be equal in measure to the matching angle of the other polygon.

example 108.1   Which two triangles appear to be similar?

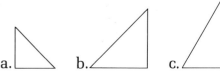

a.   b.   c.

solution   Two triangles are similar if they are the same **shape**. Triangles are the same shape if they have matching angles. The size and position of the two figures does not affect similarity. The two triangles that appear to be similar are **(a)** and **(b)**.

example 108.2   The two triangles **are** similar. What is the measure of angle A?

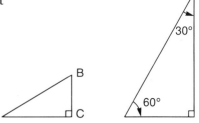

solution   Similar polygons have matching angles. Angle A is the smallest angle of the left triangle, so it must match with the smallest angle of the right triangle. Thus the measure of angle A must be **30°**.

practice   a. "All squares are similar." True or false?   true
   b. "All rectangles are similar." True or false?   false
   c. "All similar shapes are equal in size." True or false?   false
   d. "If two polygons are similar, then their matching angles arc cqual in mcasurc." Truc or falsc?   true

problem   **1.** What is the sum of the first five positive even numbers?
set 108   30

   2. The team's won-lost ratio is 4 to 3. If the team has won 12 games, how many games has the team lost?   9

   3. Thirty-six of the 88 piano keys are black. What is the ratio of black keys to white keys on a piano? $\frac{9}{13}$

   *4. Which two triangles appear to be similar?   (a), (b)

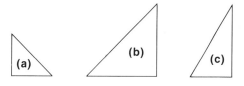

   5. Three eighths of the 48 band members played woodwinds. How many woodwind players were in the band?   18

   6. What is the least common multiple (LCM) of 6, 8, and 12?
        24

   *7. The two triangles are similar. What is the measure of angle A?   30°

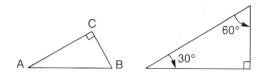

   8. What number completes the proportion? $\dfrac{3}{20} = \dfrac{n}{100}$ 15

Complete the chart to answer problems 9, 10, and 11.

	FRACTION	DECIMAL	PERCENT
9.	$1\frac{2}{5}$	**a.** 1.4	**b.** 140%
10.	**a.** $\frac{6}{25}$	0.24	**b.** 24%
11.	**a.** $\frac{7}{20}$	**b.** 0.35	35%

12. $4\dfrac{3}{4} + \left(2\dfrac{1}{4} - \dfrac{7}{8}\right) =$ $6\frac{1}{8}$ **13.** $1\dfrac{1}{5} \div \left(2 \div 1\dfrac{2}{3}\right) =$ 1

14. $6.2 + (9 - 2.79) =$ 12.41   **15.** $9 \div 1.2 =$ 7.5

16. Round the product to the nearest cent: $\$2.89 \times 0.06 =$ $0.17

17. What decimal part of a meter is a millimeter?   0.001

18. Arrange in order of size from least to greatest:
0.3, 0.31, 0.305   0.3, 0.305, 0.31

19. If each edge of a cube is 5 inches, then its **volume** is how many cubic inches?   125 cu. in.

20. $2^5 - 5^2 =$ 7   **21.** $8a = 360$ $a =$ 45

22. This acute angle is half of a right angle. Its measure is how many degrees?
45°

23. What is the perimeter of this shape?   100 mm

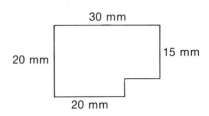

24. What is the area of the shape?   550 sq. mm

25. How long is the line segment?
$1\frac{3}{8}$ in.

LESSON
109

"Congruent"

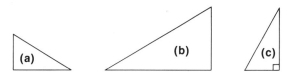

All of these triangles are similar. Two of them are **congruent**. Shapes which are similar and are also equal in size are congruent. Plane figures which are congruent can be made to fit exactly upon one another by turning, sliding, or flipping one onto the other. With the shapes above it would be possible to turn, slide, or flip figure A so that it exactly fit on figure C. We say that triangles A and C are congruent. Figures which are identical in size and shape are congruent.

example 109.1   "All figures which are congruent are also similar." True or false?

solution   Figures which are similar have the same shape or form, although they may not be the same size. Similar shapes which are the same size are also congruent. Figures which are congruent are a special case of those which are similar, so all figures which are congruent are also similar. **True.**

example 109.2   These triangles are congruent. What is the perimeter of each?

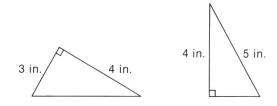

solution   Shapes which are congruent have matching parts which are equal in measure. The unlabeled side of the triangle on the left matches the 5-in. side of the triangle on the right, so the perimeter of each is 12 inches.

$$3 \text{ in} + 4 \text{ in} + 5 \text{ in} = \textbf{12 in}$$

problem set 109

1. What is the sum of the first nine prime numbers?   100

2. On the map, 2 cm equals 1 km. What is the actual length of a street which is 10 cm long on the map?   5 km

3. Between 8:00 P.M. and 9:00 P.M. the station telecast 8 minutes of commercials. What was the ratio of commercial time to program time during that hour? $\frac{2}{13}$

4. Which appears to have a right angle as one of its angles?   (c)

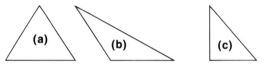

5. How many degrees would an acute angle measure if it were one-third the size of a right angle?   30°

6. What is the greatest common factor (GCF) of 10, 15, and 25?   5

7. $10^3 \div 10^2 =$ 10

8. $\dfrac{8}{n} = \dfrac{4}{25}$ $n =$ 50

Complete the chart to answer problems 9, 10, and 11.

	FRACTION	DECIMAL	PERCENT
9.	$\frac{5}{8}$	**a.** 0.625	**b.** $62\frac{1}{2}\%$
10.	**a.** $1\frac{1}{4}$	1.25	**b.** 125%
11.	**a.** $\frac{7}{10}$	**b.** 0.7	70%

12. $2\dfrac{1}{3} + \left(3\dfrac{1}{2} - 1\dfrac{2}{3}\right) = 4\dfrac{1}{6}$ **13.** $3 \times 3\dfrac{1}{3} \times 1\dfrac{1}{3} = 13\dfrac{1}{3}$

14. $(6.2 + 9) - 2.79 =$ 12.41   **15.** $4.2 \div 0.24 =$ 17.5

16. Round the product to the nearest cent: $10.80 \times 0.065 =$
$0.70

17. Write the fraction $\frac{2}{3}$ as a decimal rounded to the hundredths'
place.   0.67

18. What is the **volume** of a room which is 10 feet wide, 12
feet long, and 8 feet high? (Think of how many boxes 1
foot on each side would be needed to fill the room.)
960 cu. ft.

*19. "All figures which are congruent are also similar." True or
false?   true

20. What is $\dfrac{3}{8}$ of $3.60? $1.35 **21.** $3w - 1 = 20$ $w = 7$

22. These triangles are congruent. What
is the **area** of each?   6 sq. in.

23. What number is thirty-two less than twenty-three? -9

24. What is the area of this shape?
51 sq. cm

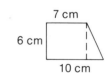

25. How many **millimeters** long is this line segment?
34 mm

LESSON
110

Classifying Triangles

All three-sided polygons are triangles, but not all triangles are alike. We distinguish between six different types of triangles. We make these distinctions based on the relative measures of the sides and on the measures of the angles.

We will first consider three different triangles based on the relative measures of their sides.

Triangles Classified by Their Sides		
Three equal sides	Equilateral	△
Two equal sides	Isosceles	◁ ◁
Three unequal sides	Scalene	◁

An **equilateral** triangle has three equal sides and three equal angles.

An **isosceles** triangle has at least two equal sides and two equal angles.

A **scalene** triangle has three unequal sides and three unequal angles.

Triangles are also classified by their angles. In an earlier lesson we learned the names of three different kinds of angles: **acute**, **right** and **obtuse**. We use these words to describe triangles as well.

Triangles Classified by Their Angles		
All acute angles	Acute triangle	△
One right angle	Right triangle	◸
One obtuse angle	Obtuse triangle	◿

practice   **a.** One side of an equilateral triangle measures 15 cm. What
(See Practice   is the perimeter of the triangle?   45 cm
Set LL in the   **b.** "An equilateral triangle is also an acute triangle." True or
Appendix.)   false?   true
                c. "All acute triangles are equilateral triangles." True or <u>false</u>?
                d. Two sides of a triangle are 3 in. and 4 in. If the perimeter
                       is 10 in., what type of triangle is it?   isosceles
                e. Is every right triangle a scalene triangle? (Yes or No)   No
                f. A scalene triangle cannot be: (a) acute (b) right (c) obtuse
                       (d) isosceles.   (d)

problem   **1.** When the greatest four-digit number is divided by the
set 110   greatest two-digit number, what is the quotient?   101

2. The ratio of the length to the width of a rectangle is 3 to
2. If the width is 60 mm, what is the length?   90 mm

3. The diameter of a certain circle is 50 cm. What is the ratio
of the length of the radius to the length of the diameter?
$\frac{1}{2}$

4. A full turn is 360°; how many degrees is $\frac{1}{6}$ of a turn?
60°

5. What is the perimeter of the equilateral triangle?
18 cm

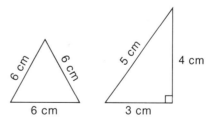

6. What is the perimeter of the scalene triangle?   12 cm

7. What type of angle is greater than an acute angle but less
than an obtuse angle?   right ∠

8. What number completes the proportion? $\frac{4}{5} = \frac{n}{100}$ 80

Complete the chart to answer problems 9, 10, and 11.

	FRACTION	DECIMAL	PERCENT
9.	$2\frac{3}{4}$	**a.** 2.75	**b.** 275%
10.	**a.** $1\frac{1}{10}$	1.1	**b.** 110%
11.	**a.** $\frac{16}{25}$	**b.** 0.64	64%

12. $26\frac{1}{6} + 23\frac{1}{3} + 22\frac{1}{2} =$ 72   **13.** $\left(1\frac{1}{5} \div 2\right) \div 1\frac{2}{3} = \frac{9}{25}$

14. $9 - (6.2 + 2.79) =$ 0.01   **15.** $63 \div 0.36 =$ 175

16. Round the product to the nearest cent: $\$24.89 \times 0.065 =$ $1.62

17. Round the quotient to the nearest thousandth: $0.065 \div 4 =$ 0.016

18. Write the prime factorization of 100. $2 \times 2 \times 5 \times 5$

19. "All squares are similar." True or false?   true

20. The two triangles shown below are congruent. If the perimeter of each is 24 cm, what is the area of each? 24 sq. cm

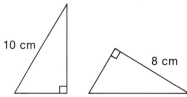

21. What is $\frac{2}{3}$ of 90?   60   **22.** $3^2 - 3^3 =$ -18

23. What is the perimeter of this shape? 44 m

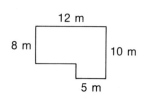

24. $12 = 2y - 2$ $y = 7$

25. How long is the line segment?
$1\frac{3}{4}$ in.

inches   1   2

LESSON
111

Writing Division Answers

When a division problem does not come out even, there are several ways to write the answer. We can write the answer with a remainder. Also, we can write the answer as a mixed number or as a decimal number.

$$
\begin{array}{c} 3\ \text{r}3 \\ 4\,\overline{)\,15} \end{array}
\qquad
\begin{array}{c} 3\frac{3}{4} \\ 4\,\overline{)\,15} \end{array}
\qquad
\begin{array}{c} 3.75 \\ 4\,\overline{)\,15.00} \end{array}
$$

Sometimes we should round an answer. How a division problem should be ended depends upon how the problem is posed.

In the problem sets which follow, we will find division problems which will require us to write our answers in various ways. Some division problems will ask us to give a reasonable answer to a particular question. In the example below, the division does not come out even. We must write our answer in a way which is reasonable.

Other division problems will ask us to practice writing answers in more than one way. The practice questions below ask us to write the answer to a division problem in three different ways.

example 111.1 One hundred students are to be assigned to 3 classrooms. How many students should be in each class so that the numbers are as balanced as possible?

solution Dividing 100 by 3 gives us $33\frac{1}{3}$ students per class, which is impractical. Assigning 33 students per class takes 99 students, so one class would have 34. We write the answer **33, 33, 34**.

practice
(See Practice
Set MM in the
Appendix.)

	REMAINDER	FRACTION	DECIMAL
$124 \div 10 =$	**a.** 12 r4	**b.** $12\frac{2}{5}$	**c.** 12.4
$93 \div 12 =$	**d.** 7 r9	**e.** $7\frac{3}{4}$	**f.** 7.75
$100 \div 16 =$	**g.** 6 r4	**h.** $6\frac{1}{4}$	**i.** 6.25

problem   *1. One hundred students are to be assigned to 3 classrooms.
set 111   How many students should be in each class so that the
              numbers are as balanced as possible? (List the numbers.)
              33, 33, 34

2. Round ten and eighty-six thousandths to the nearest
   hundredth.   10.09

3. What is three tenths of sixty?   18

4. What is the ratio of shaded to un-
   shaded circles? $\frac{3}{2}$

5. What is the standard numeral for (5 × 1000) + (4 × 10) +
   (3 × 1)?   5043

6. Write seven and one half as a decimal numeral.   7.5

7. A nickel is what percent of a dollar?   5%

8. The symbol ≠ means **not equal to**. Which is true?
   (a) 1 ≠ 100%, (b) 1 ≠ −1, (c) 1 ≠ 1.0   (b) −1

Complete the chart to answer problems 9, 10, and 11.

	FRACTION	DECIMAL	PERCENT
9.	$\frac{11}{20}$	**a.** 0.55	**b.** 55%
10.	**a.** $1\frac{1}{2}$	1.5	**b.** 150%
11.	**a.** $\frac{1}{100}$	**b.** 0.01	1%

12. $3\frac{2}{3} + 6\frac{5}{6} = 10\frac{1}{2}$ 13. $4\frac{1}{2} \div 6 = \frac{3}{4}$

14. $(4.3 + 0.41) \div 0.3 = 15.7$ 15. $6 - (0.6 - 0.06) = 5.46$

16. Using exponents, $2 \times 2 \times 2$ equals 2^3. Write $3 \times 3 \times 3 \times 3$
    using exponents. 3^4

17. How many centimeters is 1.2 meters? (1 meter = 100 cm)
120 cm

18. What is the area of this quadrilateral?
40 sq. m

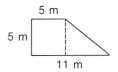

19. A pyramid with a square base has how many vertexes?
5

20. Line r is called a **line of symmetry** because it divides the equilateral triangle into two mirror images. Which other line is also a line of symmetry?   *t*

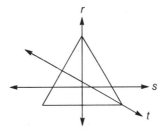

21. $3m + 1 = 10$ $m = 3$

22. Write the prime factorization of 60. $2 \cdot 2 \cdot 3 \cdot 5$

23. The candy bar has 10 sections. If one section weighs 12 grams, what does the whole candy bar weigh?   120 grams

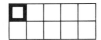

24. $24 - 50 = -26$

25. The numbers in these boxes form number patterns. What **one** number should be put in both empty boxes to complete the patterns?   6

1	2	3
2	4	
3		9

LESSON
112

Finding a Given Percent of a Number

Some questions can be asked by using a percent, by using a fraction, or by using a decimal number. These three sentences all ask the same question.

(a)   What is 25% of 20?
(b)   What is $\frac{1}{4}$ of 20?
(c)   What is 0.25 of 20?

We may answer percent questions by changing them into fraction or decimal questions, or we may think of the meaning of the percent itself.

Looking at question (a), we remember that 25% means $\frac{25}{100}$. To find $\frac{25}{100}$ of 20, we may divide 20 into 100 parts and count the value of 25 of those parts. This box shows 20 divided into 100 parts. Each part has a value of 0.2. ($20 \div 100 = 0.2$.) Twenty-five parts would be $25 \times 0.2 = 5$.

.2	.2	.2	.2	.2	.2	.2	.2	.2	.2
.2	.2	.2	.2	.2	.2	.2	.2	.2	.2
.2	.2	.2	.2	.2	.2	.2	.2	.2	.2
.2	.2	.2	.2	.2	.2	.2	.2	.2	.2
.2	.2	.2	.2	.2	.2	.2	.2	.2	.2
.2	.2	.2	.2	.2	.2	.2	.2	.2	.2
.2	.2	.2	.2	.2	.2	.2	.2	.2	.2
.2	.2	.2	.2	.2	.2	.2	.2	.2	.2
.2	.2	.2	.2	.2	.2	.2	.2	.2	.2
.2	.2	.2	.2	.2	.2	.2	.2	.2	.2

20

example 112.1   **a.** What is 1% of 60?   **b.** What is 30% of 60?

solution   **a.** To find 1 percent of 60, we divide 60 by 100.

$$1 \text{ percent of } 60 = \frac{60}{100} = \textbf{0.6}$$

b. Thirty percent of 60 is 30 times 1 percent of 60.

$$30 \text{ percent of } 60 = 30 \times 0.6 = \textbf{18}$$

practice
(See Practice
Set NN in the
Appendix.)

a. What is 1% of 20?   0.2
b. What is 50% of 20?   10
c. What is 1% of 50?   0.5
d. What is 20% of 50?   10

e. What is 1% of 25?   0.25
f. What is 65% of 25?   16.25
g. What is 1% of 140?   1.4
h. What is 30% of 140?   42

**problem
set 112**

1. Two hundred students are traveling by bus on a field trip. The maximum number of students allowed on each bus is 84. How many buses are needed for the trip?   3

2. Which is the longest distance? 3.14 m, 3.4 m, 3 m, 3.41 m
   3.41 m

*3. (a) What is 1% of 60? (b) What is 30% of 60?
   (a) 0.6   (b) 18

4. Complete the proportion: $\dfrac{6}{10} = \dfrac{9}{a}$ 15

5. In expanded notation 1760 is written as follows:
   $(1 \times 1000) + (7 \times 100) + (?)$. Find the value of "?".
   6×10

6. Write the numeral twenty million, five hundred ten thousand.   20,510,000

7. Round the product of 3.65 and 1.3 to the nearest tenth.
   4.7

8. Compare: $\dfrac{5}{6} \bigcirc \dfrac{5}{7}$ >

Complete the chart to answer problems 9, 10, and 11.

	FRACTION	DECIMAL	PERCENT
9.	$1\dfrac{4}{5}$	**a.** 1.8	**b.** 180%
10.	**a.** $\frac{3}{5}$	0.6	**b.** 60%
11.	**a.** $\frac{1}{50}$	**b.** 0.02	2%

12. $5\dfrac{1}{2} - 2\dfrac{5}{6} = 2\dfrac{2}{3}$

13. $\dfrac{2}{3} \times 2\dfrac{1}{4} \times 2 = 3$

14. $3.45 + 6.7 + 0.429 =$
10.579

15. $8 \div 0.05 = 160$

16. If three books weigh a total of 6 pounds, how much do all 8 books weigh?   16 lb.

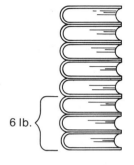

6 lb.

17. How many millimeters is 1.2 meters? (1 m = 1000 mm.)
1200 mm

18. What is the perimeter of the polygon?   54 mm

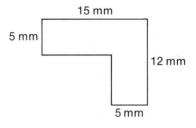

15 mm

5 mm

12 mm

5 mm

19. What type of triangle has two equal sides?   isosceles

20. If the pattern shown below were cut out and folded on the dotted lines, would it form a cube, a pyramid or a cylinder?
cube

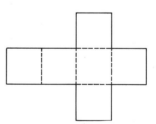

21. Solve: $\dfrac{35}{x} = \dfrac{5}{2}$ 14

22. Which is not a composite number? 34, 35, 36, 37   37

23. Four out of 10 is the same as how many out of 100?   40

24. How many years was it from 12 B.C. to 23 A.D.?   34 years

25. Which arrow points to $-\dfrac{1}{2}$?   B

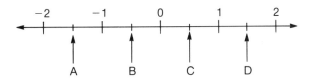

LESSON
113

Expanded Notation
Using Exponents

Numbers like 10; 100; 1000; 10,000; and so on can be named as powers of ten.

$$10^1 = 10 \qquad 10^2 = 100 \qquad 10^3 = 1000 \qquad 10^4 = 10,000$$

We can use powers of ten to name place value. Notice that 10^0 is equal to 1.

Millions			Thousands			Ones		
hundreds	tens	ones	hundreds	tens	ones	hundreds	tens	ones
10^8	10^7	10^6	10^5	10^4	10^3	10^2	10^1	10^0

In this lesson we will practice writing numbers in expanded notation using exponents.

$$5072 = (5 \times 1000) + (0 \times 100) + (7 \times 10) + (2 \times 1)$$
$$5072 = (5 \times 10^3) + (0 \times 10^2) + (7 \times 10^1) + (2 \times 10^0)$$

example 113.1   Write the number $(6 \times 10^3) + (5 \times 10^2) + (0 \times 10^1) + (3 \times 10^0)$ in standard notation.

solution   Standard notation: **6503**

example 113.2   Write 7163 in expanded notation using exponents.

solution   Using exponents: $(7 \times 10^3)+(1 \times 10^2)+(6 \times 10^1)+(3 \times 10^0)$

practice
(See Practice
Set OO in the
Appendix.)

Write these numbers in expanded notation using exponents.
a. 5030 $(5 \times 10^3) + (3 \times 10^1)$
b. 503 $(5 \times 10^2) + (3 \times 10^0)$
c. 5300 $(5 \times 10^3) + (3 \times 10^2)$

Write these numbers using standard notation.
d. $(7 \times 10^3) + (0 \times 10^2) + (3 \times 10^1) + (2 \times 10^0)$ 7032
e. $(4 \times 1000) + (6 \times 100) + (0 \times 10) + (3 \times 1)$ 4603
f. $(7 \times 10^4) + (0 \times 10^3) + (1 \times 10^2) + (3 \times 10^1)$ 70,130

**problem
set 113**

1. The winning pitcher faced 32 batters during the game. If the opposing team had only 9 players, how many of those players did the pitcher face 4 times?   5

2. If the cost of calling Tucson is $1.12 for the first 3 minutes and 52¢ for each additional minute, what would be the cost of a 10-minute call? $4.76

3. (a) What is 1% of 30? (b) What is 90% of 30?
(a) 0.3   (b) 27

4. Six of the 18 lights were on. What is the ratio of lights on to lights off? $\frac{1}{2}$

5. Write 5300 in expanded notation using exponents.
$(5 \times 10^3) + (3 \times 10^2)$

6. Write the decimal number twelve and twenty-four thousandths.   12.024

7. Estimate the product of 4.92 and 12.1 to the nearest whole number.   60

8. Arrange in order from greatest to least: 0.2, -2, $\frac{1}{2}$
$\frac{1}{2}$, 0.2, -2

Complete the chart to answer problems 9, 10, and 11.

	FRACTION	DECIMAL	PERCENT
9.	$\dfrac{3}{25}$	**a.** 0.12	**b.** 12%
10.	**a.** $\frac{7}{20}$	0.35	**b.** 35%
11.	**a.** $\frac{1}{25}$	**b.** 0.04	4%

12. $\left(3\dfrac{1}{2} + 2\dfrac{1}{3}\right) - 1\dfrac{3}{4} = 4\dfrac{1}{12}$ **13.** $1\dfrac{1}{3} \div 2\dfrac{2}{3} = \dfrac{1}{2}$

14. $0.1 \times 2.6 \times 0.07 = $ 0.0182   **15.** $(1.2 - 0.24) \div 3 = $ 0.32

16. Using exponents, $3 \times 3 \times 4 \times 4 \times 4 = 3^2 \times 4^3$. Write $3 \times 3 \times 3 \times 4 \times 4$ using exponents. $3^3 \times 4^2$

17. How many millimeters is 62.3 cm? (1 cm = 10 mm.)
623 mm

18. What is the area of the hexagon shown here?
1700 sq. mm

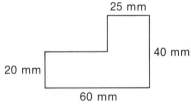

25 mm

40 mm

20 mm

60 mm

19. What is the volume of a cube with edges 10 cm long?
1000 cu. cm

20. A line of symmetry divides a figure into 2 mirror images. Which line is **not** a line of symmetry in this figure?   r

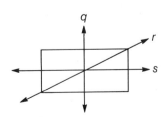

21. $3 \times (4 + 5) = (3 \times 4) + (m \times 5)$ $m = 3$

22. Make a list of the prime numbers that are even.   2

23. The money was separated into 10 equal parts. The money in 4 of the parts totaled $12. How much money was in all 10 parts? $30

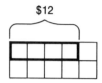

$12

24. $26 - 62 = -36$

25. What type of triangle has three unequal sides?
scalene

LESSON
114

Multiplying by Powers of Ten

We can multiply by powers of ten very easily. Multiplying by powers of ten does not change the digits, only the place values of the digits. We can change the place values by moving the decimal point the number of places shown by the exponent. To write 1.2×10^3 in standard notation we simply move the decimal point three places to the right and fill the empty places with zero: $1.2 \times 10^3 = 1.200 = 1200$.

example 114.1   Write 6.2×10^2 in standard notation.

solution   To multiply by a power of ten, simply move the decimal point the number of places shown by the exponent. In this case move the decimal point two places to the right. $6.2 \times 10^2 = 6.20 = $ **620**.

Sometimes powers of ten are named with words instead of numerals. For example, we might read that there are 5.2 million people living in Hong Kong. The numeral 5.2 million means $5.2 \times 1,000,000$. We can write this number by shifting the decimal point six places to the right, which gives us 5,200,000.

practice Write the standard notation for each of the following numbers. Always begin by changing fractions and mixed numbers to decimal numbers.

a. 1.2×10^4
12,000

b. 3×10^3
3000

c. 4.2×10^2
420

d. 1.23×10^1
12.3

e. 11×10^3
11,000

f. 23 million
23,000,000

g. 1.5 million
1,500,000

h. 5 billion
5,000,000,000

i. 4.2 billion
4,200,000,000

j. 0.5 million
500,000

k. $\frac{1}{2}$ million
500,000

l. $\frac{1}{4}$ billion
250,000,000

m. $\frac{3}{4}$ million
750,000

n. $1\frac{1}{2}$ billion
1,500,000,000

o. $2\frac{1}{4}$ million
2,250,000

problem set 114

1. Eight people are to have equal shares of one dozen ice cream sandwiches. How many sandwiches should each person receive? (The sandwiches can be broken into pieces as required.) $1\frac{1}{2}$

2. It is $\frac{3}{4}$ of a mile from Mark's house to school. How far would Mark walk going to school and back for 5 days? $7\frac{1}{2}$ mi.

3. Eighty percent of the 30 students passed the test. How many students passed the test? 24

4. Complete the proportion: $\frac{12}{25} = \frac{y}{100}$ 48

5. Write the standard numeral for $(6 \times 10^4) + (2 \times 10^2)$.
60,200

6. Write the numeral for ten million, one hundred one thousand. 10,101,000

7. What is the average of 6.23, 4.39, and 7.2? 5.94

8. Which of the following numbers is closest to 1? 1.11, 0.90, 0.91 0.91

Complete the chart to answer problems 9, 10, and 11.

	FRACTION	DECIMAL	PERCENT
9.	$2\frac{1}{4}$	**a.** 2.25	**b.** 225%
10.	**a.** $\frac{2}{25}$	0.08	**b.** 8%
11.	**a.** $\frac{1}{20}$	**b.** 0.05	5%

12. $3\frac{1}{2} + 6\frac{2}{3} + 2\frac{5}{6} =$ 13

13. $7\frac{1}{2} \div 1\frac{1}{2} =$ 5

14. $(3.2 + 12) \times 0.15 =$ ·2.28

15. $(12 - 3.6) \div 0.12 =$ 70

16. Using exponents $2 \cdot 2 \cdot 2 \cdot 3 \cdot 3 = 2^3 \cdot 3^2$. Write $2 \cdot 2 \cdot 3 \cdot 3 \cdot 3$ using exponents. $2^2 \cdot 3^3$

17. Ten kilometers is how many meters? (1 km = 1000 m.)
10,000 m

18. What is the area of this parallelogram?
72 sq. m

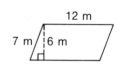

19. A pyramid with a triangular base has how many faces?
4

20. How many degrees is $\frac{1}{4}$ of a straight angle?   45°

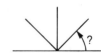

21. What is the greatest common factor (GCF) of 40 and 100?
20

22. The boys were divided into 8 equal teams. Fourteen boys made 2 teams. How many boys were there in all?
56

*23. Write 6.2×10^2 as a standard numeral.   620

24. What number is missing in the table?   5

n	5	4	3	2
$2n - 1$	9	7	?	3

25. To what decimal number is the arrow pointing?
4.4

LESSON
115

Writing Rates as Ratios

Ratios are used in a number of ways. One common type of ratio is **rate**. Speed, batting average, bank interest, taxes, and wages are all rates. At this level we will practice rate problems which can be solved by multiplying.

First we will practice writing rates as ratios. See how the following rates have been written as ratios. Notice how the units are written in the ratio. Notice also that the words **per**, **in**, **for each**, and **out of** indicate rates and are written with a division or fraction line.

examples and solutions

55 miles per hour $\dfrac{55 \text{ miles}}{1 \text{ hour}}$

3 hits in 10 at-bats $\dfrac{3 \text{ hits}}{10 \text{ at-bats}}$

6 cents for each dollar $\dfrac{6 \text{ cents}}{1 \text{ dollar}}$

9 out of 10 Americans are milk drinkers $\dfrac{9 \text{ milk drinkers}}{10 \text{ Americans}}$

practice Use fractions or whole numbers as required to write these rates as ratios.

a. 3 dollars per hour
$\dfrac{\$3.00}{1 \text{ hour}}$

b. 6 baskets out of 10 shots
$\dfrac{6 \text{ baskets}}{10 \text{ shots}}$

c. 10 cents per kilowatt hour
$\dfrac{10 \text{ cents}}{1 \text{ kilowatt hour}}$

d. 29 students per teacher
$\dfrac{29 \text{ students}}{1 \text{ teacher}}$

e. 4 of the 25 answers were wrong
$\dfrac{4 \text{ wrong answers}}{25 \text{ answers}}$

f. 160 kilometers in 2 hours
$\dfrac{160 \text{ kilometers}}{2 \text{ hours}}$

g. 26 miles per gallon
$\dfrac{26 \text{ miles}}{1 \text{ gallon}}$

h. 12 inches for each foot
$\dfrac{12 \text{ in.}}{1 \text{ ft.}}$

i. $0.67 per 100 cubic feet
$\dfrac{\$0.67}{100 \text{ cu. ft.}}$

j. 2500 revolutions per minute
$\dfrac{2500 \text{ revolutions}}{1 \text{ min.}}$

problem set 115

1. For cleaning the yard the four boys were paid a total of $15. How much is each boy's fair share? $3.75

2. About how many meters long is a bicycle?   1, 2, 6, 36
2 m

3. Eighty percent of the air is nitrogen. What percent is not nitrogen? 20%

*4. Write the ratio for 3 hits in 10 at-bats. $\dfrac{3 \text{ hits}}{10 \text{ at-bats}}$

5. Write the standard numeral for $(6 \times 10^3) + (2 \times 10^1) + (4 \times 10^0)$. 6024

6. Use a decimal numeral to write twenty and five hundred eight thousandths. 20.508

7. Round the quotient of $1.4 \div 0.3$ to the nearest hundredth. 4.67

8. Arrange in order from least to greatest: $\dfrac{3}{4}, \dfrac{3}{5}, \dfrac{4}{5}$

$\frac{3}{5}, \frac{3}{4}, \frac{4}{5}$

Complete the chart to answer problems 9, 10, and 11.

	FRACTION	DECIMAL	PERCENT
9.	$\dfrac{1}{50}$	**a.** 0.02	**b.** 2%
10.	**a.** $1\frac{3}{4}$	1.75	**b.** 175%
11.	**a.** $\frac{1}{4}$	**b.** 0.25	25%

12. $12\dfrac{1}{4} - 3\dfrac{5}{8} = 8\frac{5}{8}$

13. $3\dfrac{1}{3} \times 2\dfrac{1}{4} \times 1\dfrac{3}{5} = 12$

14. $12 + 7.65 + 15.8 = 35.45$

15. $0.59 \times 0.17 = 0.1003$

16. $5^2 + 2^5 = 57$

17. What is the area of the trapezoid?
34 sq. ft.

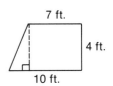

7 ft.

4 ft.

10 ft.

18. How many inches is 1 yard, 2 feet, 3 inches? (1 yd. = 36 in., 1 ft. = 12 in.) 63 in.

19. Which line in this figure is not a line of symmetry?

g

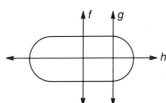

20. How many cubes one centimeter on each edge would be needed to fill this box?   30 cu. cm

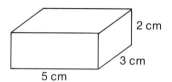

21. John worked for 10 days. After 3 days he had earned $60. How much did he earn after 10 days? $200

22. Seventy is the product of which three prime numbers?
$2 \times 5 \times 7$

23. Saturn is 900 million miles from the sun. Write the number in expanded notation using exponents. 9×10^8

24. To which number is the arrow pointing?   40

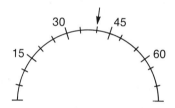

25. The ratio of quarters to dimes in the soda machine was 5 to 8. If there were 120 quarters, how many dimes were there?   192

LESSON
116

Unit Canceling

We are getting ready to practice rate problems which can be solved by multiplying. An important idea to have in mind when multiplying rates is that we can cancel units just like we cancel numbers. For instance, we may see a problem where miles per hour and hours are multiplied together. The units cancel as shown on the next page.

$$\frac{4 \text{ miles}}{1 \text{ hour}} \times \frac{2 \text{ hours}}{1} = 8 \text{ miles}$$

Notice that we write 2 hours as 2 hours over 1.

example 116.1 $\dfrac{55 \text{ miles}}{1 \text{ hour}} \times \dfrac{6 \text{ hours}}{1} =$

solution Cancel units and multiply. $\dfrac{55 \text{ miles}}{1 \text{ hour}} \times \dfrac{6 \text{ hours}}{1} = \textbf{330 miles}$

practice Multiply these amounts, canceling both numbers and units whenever possible.

a. $\dfrac{3 \text{ dollars}}{1 \text{ hour}} \times \dfrac{8 \text{ hours}}{1} =$ 24 dollars

b. $\dfrac{6 \text{ baskets}}{10 \text{ shots}} \times \dfrac{100 \text{ shots}}{1} =$ 60 baskets

c. $\dfrac{10 \text{ cents}}{1 \text{ kwh}} \times \dfrac{26.3 \text{ kwh}}{1} =$ 263 cents

d. $\dfrac{29 \text{ students}}{1 \text{ teacher}} \times \dfrac{18 \text{ teachers}}{1} =$ 522 students

e. $\dfrac{160 \text{ km}}{2 \text{ hours}} \times \dfrac{10 \text{ hours}}{1} =$ 800 km

f. $\dfrac{2.3 \text{ m}}{1} \times \dfrac{100 \text{ cm}}{1 \text{ m}} =$ 230 cm

problem set 116

1. Tickets to the movie are \$3 each. How many tickets can Jan buy with \$20? 6

2. Mary ran four laps of the track at an even pace. If it took 6 minutes to run the first three laps, how long did it take to run all four laps? 8 min.

3. Fifteen of the 25 members played. What fraction of the members did not play? $\frac{2}{5}$

4. Compare: $\dfrac{5}{6} \bigcirc \dfrac{11}{13}$ <

5. Which digit in 94,763,581 is in the ten-thousands' place?
6

6. What number is ten more than ninety-nine thousand, nine hundred ninety?   100,000

7. Estimate the sum of $36.43, $41.92, and $26.70 to the nearest dollar. $105

8. Which of these numbers is closest to 10? $\frac{19}{2}$, 9.2, 10.9
$\frac{19}{2}$

Complete the chart to answer problems 9, 10, and 11.

	FRACTION	DECIMAL	PERCENT
		a.	**b.**
9.	$\frac{1}{8}$	0.125	$12\frac{1}{2}\%$
10.	**a.** $\frac{9}{10}$	0.9	**b.** 90%
11.	**a.** $\frac{3}{5}$	**b.** 0.6	60%

12. $3\frac{1}{4} + 2\frac{1}{2} + 4\frac{5}{8} = 10\frac{3}{8}$

13. $3.25 \div \frac{2}{3} =$ (fraction answer) $4\frac{7}{8}$

14. $5 - 1.375 = 3.625$ **15.** Solve: $\frac{3}{2} = \frac{18}{m}$ 12

16. Write $3 \cdot 3 \cdot 3 \cdot 3 \cdot 4 \cdot 4 \cdot 4$ using exponents. $3^4 \cdot 4^3$

*17. Cancel units and multiply: $\frac{55 \text{ miles}}{1 \text{ hour}} \times \frac{6 \text{ hours}}{1} = 330 \text{ mi.}$

18. What is the area of the polygon on the next page?
5.52 sq. m

19. What is the perimeter of this polygon?   10 m

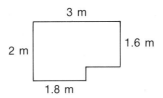

20. $5w - 1 = 49$ $w = 10$

21. Which two lines in this figure appear to be parallel?
q, r

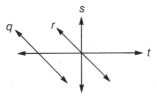

22. Which two lines appear to be perpendicular?   s, t

23. What is the least common multiple (LCM) of 12 and 8?
24

24. Write 1.5×10^6 as a standard numeral.   1,500,000

25. There are two ways of making 5¢: with five pennies or with one nickel. How many ways can we use coins to make 12¢?   4

LESSON 117

Rounding Numbers— A Three-Step Method

We have practiced rounding numbers with the idea that we are looking for the multiple of 10, 100, 1000, etc.—to which the number we are rounding is closest. When rounding 257 to the nearest ten, we considered that 257 is between 250 and 260 and that it is closer to 260. In this lesson we will learn a three-step method for rounding, and we will practice rounding numbers with more places. To illustrate this method we will round 15,876,285 to the ten-thousands' place.

1. Mark the place to which the number is to be rounded with a circle and write an arrow over the following place.

$$1\,5\,,8\,\overset{\downarrow}{\textcircled{7}}\,6\,,2\,8\,5$$

2. Change the arrow-marked digit and all digits to its right to zero.

$$1\,5\,,8\,\overset{\downarrow}{\textcircled{7}}\,0\,,0\,0\,0$$

3. If the arrow-marked digit was 5 or more, increase the circled digit by 1. If the arrow-marked digit was less than 5, leave the circled digit unchanged.

$$1\,5\,,8\,8\,0\,,0\,0\,0$$

example 117.1   Round 378,495 to the nearest thousand.

solution   We mark the thousands' place with a circle and the following place with an arrow.

$$3\,7\,\textcircled{8}\,,\overset{\downarrow}{4}\,9\,5$$

We change the arrow-marked digit and all digits to its right to zero.

$$3\,7\,\textcircled{8}\,,\overset{\downarrow}{0}\,0\,0$$

Since the arrow-marked digit was less than 5, we leave the circled digit unchanged.

3 7 8 , 0 0 0

example 117.2   Round 42,984,135 to the nearest hundred thousand.

solution   Mark with a circle and an arrow.

$$4\,2\,,\overset{\downarrow}{\textcircled{9}}\,8\,4\,,1\,3\,5$$

Change the arrow-marked digit and all digits to its right to zero.

$$4\,2\,,\overset{\downarrow}{\textcircled{9}}\,0\,0\,,0\,0\,0$$

Increase the circled digit by 1 if the arrow marked digit was 5 or more.

4 3 , 0 0 0 , 0 0 0

Notice that when the digit 9 is rounded up, it affects other places to its left.

practice   **a.** Round 129,375,403 to the nearest ten million.   130,000,000
  b. Round 7,896,587 to the ten-thousands' place.   7,900,000
  c. Round 38,998,764 to the nearest hundred thousand.
      39,000,000

problem **1.** The outside walls of Mike's house have a surface area of
set 117   1500 square feet. He wants to paint the walls with a paint
      which covers 400 square feet per gallon. How many 1-gallon
      cans of paint should he buy?   4

  2. At the price of 4 for a dollar, what would be the price of 10?
      $2.50

  3. How far will a mallard duck fly in 12 hours if it flies at a
      rate of 24 miles per hour?   288 mi.

 *4. Round 378,495 to the nearest thousand.   378,000

  5. Write 6010 in expanded notation using exponents.
      $(6 \times 10^3) + (1 \times 10^1)$

 *6. Round 42,984,135 to the nearest hundred thousand.
      43,000,000

  7. If 10 numbers have a sum of 62, what is their average?
      6.2

  8. The mixed number $8\frac{5}{9}$ is closest to which whole number?
      9

Complete the chart to answer problems 9, 10, and 11.

	FRACTION	DECIMAL	PERCENT
9.	$\frac{7}{8}$	**a.** 0.875	**b.** $87\frac{1}{2}\%$
10.	**a.** $\frac{12}{25}$	0.48	**b.** 48%
11.	**a.** $1\frac{1}{4}$	**b.** 1.25	125%

12. $\left(5 - 1\frac{2}{3}\right) - 2\frac{1}{2} = \frac{5}{6}$ **13.** $\left(3 \times 1\frac{1}{2}\right) \div 1\frac{2}{3} = 2\frac{7}{10}$

14. $(3.45 - 0.6) \div 30 = 0.095$ **15.** $0.375 \times 0.016 = 0.006$

16. Write $2 \times 2 \times 2 \times 2 \times 3 \times 3$ using exponents.
$2^4 \times 3^2$

17. Cancel units and multiply: $\dfrac{3 \text{ ft.}}{1 \text{ yd.}} \times \dfrac{5 \text{ yd.}}{1} = 15 \text{ ft.}$

18. Which lines are lines of symmetry for the square? r, s, t

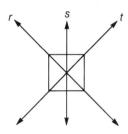

19. $50 \div 1 = 50 - w$ $w = 0$ **20.** $10 - 100 = -90$

21. Which angle appears to have a measure of 45°? c

22. Which number is missing in this sequence? 11, 15, ___, 23, 27   19

23. How many sugar cubes would be needed to build this prism?   54

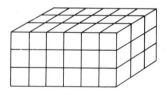

24. Write the standard numeral for 2.1 million.   2,100,000

25. Approximately how far apart are Belmond and Clear Lake?
30 mi.

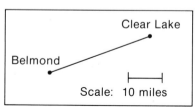

Clear Lake

Belmond

Scale: 10 miles

LESSON
118

What Percent?

Pictures can often help us work a problem. A picture can help us answer these questions.

Four is what fraction of 8?
Four is what decimal part of 8?
Four is what percent of 8?

In each question there are 8 in the whole group. Four is a part of the group.

We are asked to name **what part** of the group. Parts can be named as common fractions, as decimal fractions, and as percents. Here $\frac{4}{8}$ of the group is named. The fraction $\frac{4}{8}$ reduces to $\frac{1}{2}$ and converts to the decimal 0.5 and to 50%.

The steps are: picture the question, name the fraction, then reduce or convert to a decimal or to a percent as required.

example 118.1   Six is what percent of 8?

solution   Six is part of 8. We are to name which part as a percent. First we picture the problem. We name the part as a fraction, $\frac{6}{8}$. We write $\frac{6}{8}$ as a percent:

$$\frac{6}{8} \times \frac{100\%}{1} = \frac{600\%}{8} = \mathbf{75\%}$$

practice
(See Practice
Set PP in the
Appendix.)

a. Two is what percent of 8?   25%

b. Two is what percent of 5?   40%

c. Five is what percent of 10?   50%

d. What percent of 10 is 4?   40%

e. What percent of 6 is 2? $33\frac{1}{3}\%$

f. Six is what fraction of 10? $\frac{3}{5}$

g. What decimal part of 5 is 3?   0.6

h. What percent of 5 is 4?   80%

i. What fraction of 9 is 6? $\frac{2}{3}$

j. What decimal part of 10 is 9?   0.9

**problem
set 118**

1. Mrs. Barker wants to cut a 30-inch-long piece of licorice into 4 equal pieces. How long should she cut each piece? $7\frac{1}{2}$ in.

2. It cost $8.91 to replace 3 panes of glass. How much would it cost to replace 5 panes? $14.85

*3. Six is what percent of 8? 75%

4. A dolphin swims 24 miles per hour. How far could it travel in $1\frac{1}{2}$ hours at that rate? 36 mi.

5. Which digit is in the ten-thousandths' place in 6.73428? 2

6. Round five hundred six million, nineteen thousand to the nearest hundred thousand. 506,000,000

7. If the average of 5 numbers is 24, what is their sum? 120

8. In size, which number is between the other two? 7.1
$\frac{20}{3}$, 7.1, $7\frac{1}{2}$

Complete the chart to answer problems 9, 10, and 11.

	FRACTION	DECIMAL	PERCENT
9.	$1\frac{1}{5}$	**a.** 1.2	**b.** 120%
10.	**a.** $\frac{1}{20}$	0.05	**b.** 5%
11.	**a.** $2\frac{1}{2}$	**b.** 2.5	250%

12. $3\frac{1}{3} + 2\frac{1}{2} + 6\frac{3}{4} = 12\frac{7}{12}$ 13. $6\frac{2}{3} \div 1\frac{2}{3} = 4$

14. $(3 + 2.16) \div 0.6 = 8.6$ 15. Solve: $\frac{2}{p} = \frac{10}{15}$ 3

16. $4^3 \div 2^4 = 4$

17. Cancel units and multiply: $\dfrac{60 \text{ sec.}}{1 \text{ min}} \times \dfrac{60 \text{ min.}}{1 \text{ hr.}} \times \dfrac{24 \text{ hr.}}{1} =$
86,400 sec.

The two triangles shown are congruent.

18. What is the perimeter of each triangle?   24 cm

19. What is the area of each triangle?   24 sq. cm

20. If the two triangles were put together to make a parallelogram, what would be the area of the parallelogram?
48 sq. cm

21. In 6 years Tom will be 21. How old was he 6 years ago?
9 years

22. Write the prime factorization of 84. $2 \cdot 2 \cdot 3 \cdot 7$

23. Write the standard numeral for 3×10^5.   300,000

24. During the day the temperature rose from $-5°$ F to $12°$ F. How many degrees did the temperature rise? $17°$ F

25. To which fraction is the arrow pointing? $\dfrac{5}{6}$

LESSON

119

Circle Measure—
Circumference (Project)

Recall that the distance around a circle is its **circumference** and the distance across a circle is its **diameter**. The circumference of a circle is related to the diameter of a circle in a special way.

project   Select a circular object such as a plate or the top of a can. Use a cloth tape measure to carefully measure the circumference and diameter of the object. We recommend measuring in millimeters or centimeters. If a cloth tape measure is not available, a meterstick may be used by carefully rolling the object one full turn along the meterstick. Then divide the measure of the circumference by the measure of the diameter. You may want to use a calculator to perform the division.

$$\frac{\text{Circumference}}{\text{Diameter}} = \text{the number of diameters in the circumference}$$

How many diameters equal a circumference? 2? 3? 4? Some number in between? Compare your answer with the answers of others in the class.

practice   A circle has a diameter of 10 inches. Use the result of your investigation to estimate the circumference of this circle.
Approximately 31 inches

problem set 119

1. The car traveled 351 miles on 15 gallons of gasoline. The car averaged how many miles per gallon? (Write a decimal answer.) $23.4 \frac{\text{mi.}}{\text{gal.}}$

2. Which is the greatest weight? 6.24 lb., 6.4 lb., 6.345 lb.
6.4 lb.

3. Six is what percent of 10?   60%

4. Sound travels about 331 meters per second in air. How far will it travel in 60 seconds?   19,860 m

5. Write the standard numeral for $(5 \times 10^4) + (6 \times 10^2)$.
50,600

6. If the radius of a circle is seventy-five hundredths, what is the diameter?   1.5

7. Round the product of $3\frac{2}{3}$ and $2\frac{1}{3}$ to the nearest whole number.   9

8. Compare: 3.71 ◯ 3.709   >

Complete the chart to answer problems 9 and 10.

	FRACTION	DECIMAL	PERCENT
9.	$2\frac{2}{5}$	**a.** 2.4	**b.** 240%
10.	**a.** $\frac{17}{20}$	0.85	**b.** 85%

11. a. What is 1% of 622?   6.22
b. What is 30% of 622?   186.6

12. Solve: $\dfrac{x}{7} = \dfrac{35}{5}$ 49

13. $12\frac{1}{2} \div 1\frac{1}{9} = 11\frac{1}{4}$

14. $3.62 + 12 + 16.9 =$ 32.52

15. $0.12 \div (12 \div 0.4) =$ 0.004

16. Write $\dfrac{22}{7}$ as a decimal rounded to the hundredths' place.
3.14

17. What number multiplied by itself equals 100?   10

18. What is the area of this hexagon?
0.68 sq. ft.

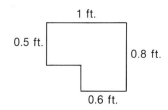

19. We call the distance across a circle the diameter. What do we call the distance around a circle?   circumference

20. What is the volume of this cube?
27 cu. cm

3 cm

21. What number is ten more than half of twelve?   16

22. If 7 containers can hold 84 ounces, how much can 10 containers hold?
120 oz.

84 ounces

23. Write the standard numeral for $4\frac{1}{2}$ million.
4,500,000

24. Round 58,697,284 to the nearest ten thousand.
58,700,000

25. Which arrow is pointing to 0.4?   C

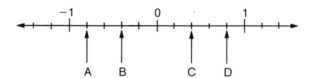

LESSON
120

Pi (π)

In the last lesson we asked the question, "How many diameters equal a circumference?" The answer to that question is **pi** (pronounced like **pie**). Pi is the name of a Greek letter which is written π. The reason the answer to the question is pi is because we are unable to write a numeral which exactly answers the question. We are only able to write numerals which are close to the value of pi.

    If you experimented to find the number of diameters in a circumference in the last lesson, you probably found that a circumference is a little more than 3 diameters. The answers vary because we are not able to measure exactly, but there **is** an exact number of diameters in the circumference of any circle, and that number of diameters is π. The number π is close to the decimal number 3.14 and to the fraction $3\frac{1}{7}$ (often written $\frac{22}{7}$). We can use these numbers to find the approximate circumference of a circle with a known diameter.

example 120.1   The diameter of a circle is 10 cm. What is the circumference? (Use $\pi = 3.14$.)

    solution   The circumference of a circle is π times the diameter. We are told to use 3.14 for π. 3.14 × 10 cm = **31.4 cm**.

practice Use π = 3.14 to find the circumference of circles with the following diameters.

    a. 100 cm   314 cm

    b. 20 in.   62.8 in.

    c. 4 ft.   12.56 ft.

    d. 1 m   3.14 m

    e. 1000 mm   3140 mm

    f. 6 miles   18.84 mi.

problem set 120

1. The 306 students were assigned to rooms so that there were 30 or 31 students in each room. How many rooms had exactly 30 students?   4

2. If 5 feet of ribbon cost $1.20, then 6 feet of ribbon would cost how much? $1.44

3. Six is what percent of 15?   40%

4. If the sales tax rate is 6 cents tax per $1 in sales, then what is the tax on a $12 sale?   72 cents

5. If $\frac{2}{5}$ of the 30 students are boys then what is the boy-girl ratio? $\frac{2}{3}$

6. Write 1.2×10^9 as a standard numeral.   1,200,000,000

7. Estimate the product of $3\frac{2}{3}$ and $2\frac{1}{3}$ to the nearest whole number.   8

8. Arrange in order from least to greatest: 9.9, 9.95, 9.925
   9.9, 9.925, 9.95

Complete the chart to answer problems 9 and 10.

	FRACTION	DECIMAL	PERCENT
9.	$3\frac{3}{8}$	**a.** 3.375	**b.** $337\frac{1}{2}\%$
10.	**a.** $\frac{3}{20}$	**b.** 0.15	15%

11. **a.** What is 1% of 9320?   93.2

    b. What is 40% of 9320?   3728

12. Solve: $\dfrac{x}{3} = \dfrac{16}{12}$ 4

13. $1\dfrac{1}{2} \times 1\dfrac{2}{3} \times 3\dfrac{1}{5} = $ 8

14. $6 + 3\dfrac{3}{4} + 4.625 = $ (decimal answer)   14.375

15. $0.67 \times 0.18 \times 1000 = $ 120.6

16. Write $3 \cdot 2 \cdot 3 \cdot 2 \cdot 3 \cdot 2 \cdot 3$ using exponents. $2^3 \cdot 3^4$

17. If a 32-ounce box of cereal costs $1.92, what is the cost per ounce?   6¢

18. What is the area of this polygon?
28 sq. m

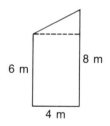

*19. The diameter of a circle is 10 cm. What is the circumference? (Use $\pi = 3.14$.)   31.4 cm

20. The volume of the pyramid is $\frac{1}{3}$ the volume of the cube. What is the volume of the pyramid?   9 cu. cm

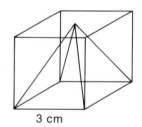

21. $6y + 2 = 20$ $y = $ 3

22. What is the greatest common factor (GCF) of 12 and 24?
12

23. What type of a triangle contains an angle which measures 90°?   right

24. What number is 4 more than -1?   3

25. What temperature is shown on the thermometer? $-2°$ C

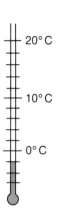

Adding Integers

Integers are the whole numbers and their negatives. Thus integers may be positive, negative, or zero. The dots on the number line above mark the integers from -5 through $+5$. We can add, subtract, multiply, and divide integers. In this lesson we will learn a game which will help us add integers.

One way to understand the addition of integers is to think of positive and negative amounts battling each other. In the game of "Sign Wars," we want to find out **who survives?**

In this battle there are 2 positives battling 3 negatives. The negatives will win because there are more of them, but the positives in losing will destroy 2 negatives. Who survives? **One negative**.

+	−
	−
+	−

This time 3 negatives are battling 7 positives. Who will win? How many will survive? **+4**.

−3
+7

example 121.1   Add: $(+8) + (-5) =$

solution   We may think of $+8$ and -5 as forces battling each other, 8 positives against 5 negatives. Who survives? The positives win because there are more of them, but the negatives in losing will destroy 5 positives, leaving **+3**.

example 121.2   Add: $(-5) + (-3) =$

solution   In this case there is no battle since both amounts are negative. Who survives? All 8 negatives survive: **−8**.

example 121.3   Add: $(-6) + (+6) =$

solution   In this battle there are the same number of positives as negatives. All are destroyed. There are no survivors. For our answer we write zero: **0**.

practice   **a.** $(-2) + (+3) = +1$ **d.** $(-3) + (+2) = -1$
(See Practice   **b.** $(+6) + (-10) = -4$ **e.** $(-5) + (+2) = -3$
Set QQ in the   **c.** $(-4) + (-6) = -10$ **f.** $(+3) + (-3) = 0$
Appendix.)

**problem
set 121**

1. If Tony reads 30 pages each day, how many days will it take him to finish a 200-page book?   7 days

2. What fraction of the letters in the word **Alabama** are A's? $\frac{4}{7}$

3. Six is what percent of 9? $66\frac{2}{3}\%$

4. If Sarah is batting at a rate of 3 hits in every 10 at-bats, then how many hits is she likely to get in 80 at-bats?   24

5. Write the numeral ninety-six million, fifty thousand, eight hundred.   96,050,800

6. $(10 + 9) + 8 = 10 + (r + 8)$ $r = 9$

7. The average of 3 numbers is 7. Two of the numbers are 8 and 9. What is the third?   4

*8. $(+8) + (-5) = +3$

9. **a.** What is 1% of 643?   6.43
   b. What is 30% of 643?   192.9

Complete the chart to answer problems 10 and 11.

	FRACTION	DECIMAL	PERCENT
10.	**a.** $1\frac{3}{10}$	1.3	**b.** 130%
11.	**a.** $\frac{3}{50}$	**b.** 0.06	6%

12. $6 - \left(4\frac{1}{4} - 1\frac{1}{2}\right) = 3\frac{1}{4}$

13. $5 \div 1\frac{2}{3} = 3$

14. $5.1 - 4\frac{1}{8} =$ (decimal answer)   0.975

15. $4 \div (0.24 \div 6) =$ 100

*16. $(-5) + (-3) = -8$

17. What is the area of this parallelogram?   3.2 sq. m

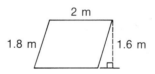

18. One hour and ten minutes is how many seconds?
4200 sec.

19. The diameter of a circle is 8 in. What is its circumference?
$(\pi = 3.14.)$ 25.12 in.

20. Complete the proportion: $\dfrac{25}{30} = \dfrac{w}{24}$ 20

21. What type of triangle has three angles smaller than 90°?
acute

22. If 12 mongooses weigh 72 pounds, how much would 100 mongooses weigh?   600 lb.

23. Pluto's average distance from the sun is about $3\frac{1}{2}$ billion miles. Write that number.   3,500,000,000

*24. $(-6) + (+6) = 0$

25. What is the cost of 3 hamburgers and 2 large drinks?
$5.65

Hamburger	$1.35
Fries	0.80
Drinks: Sm.	0.60
Lg.	0.80

LESSON
122

Finding the Whole When a Fraction Is Known

If the sports section of the newspaper is 20 pages long and if the sports section is $\frac{1}{5}$ of the newspaper, can we figure out how many pages are in the newspaper?

Drawing pictures can help us understand questions like these. The rectangle represents the whole newspaper which has been divided into 5 equal parts. There are 20 pages in each part. Since there are 20 pages in each of the 5 parts there must be 5 times 20, or 100 pages, in the whole newspaper.

Whole

20

example 122.1 Three eighths of the people in the town voted. If 120 of the people in the town voted, how many people lived in the town?

solution We are told that $\frac{3}{8}$ of the town voted, so we divide the whole into 8 parts and mark off 3 of the parts. We are told that these 3 parts total 120 people. If the 3 parts are 120, then each part must be 40 (120 ÷ 3 = 40). If each part is 40, then all 8 parts must be 8 × 40 = **320 people**.

Town

120 { (40) (40) (40)

practice
(See Practice
Set SS in the
Appendix.)

Draw a picture to help work each one of these problems.

a. Five is $\frac{1}{4}$ of what?   20

b. Six is $\frac{1}{3}$ of what?   18

c. Eight is $\frac{1}{10}$ of what?   80

d. Ten is $\frac{1}{5}$ of what?   50

e. Nine is $\frac{3}{4}$ of what?   12

f. Twenty is $\frac{2}{5}$ of what?   50

g. Sixty is $\frac{3}{8}$ of what?   160

h. Twenty-four is $\frac{3}{10}$ of what?
80

**problem
set 122**

*1. Three eighths of the people in the town voted. If 120 of the people in the town voted, how many people lived in the town?   320

2. If 130 children are separated as equally as possible into 4 groups, how many will be in each group? (List 4 answers.)
32, 32, 33, 33

3. If the parking lot charges 25¢ per half hour, what is the cost of parking a car from 11:15 A.M. to 2:45 P.M.? $1.75

4. What percent of this circle is shaded?
25%

5. Four out of 50 is the same as how many out of 100?   8

6. At 4 miles per hour how far would a person walk in $4\frac{1}{2}$ hours?   18 mi.

7. Write one hundred five thousandths as a decimal numeral.
0.105

8. Round the quotient of $7.00 ÷ 9 to the nearest cent.
$0.78

9. Arrange in order from least to greatest: 81%, $\frac{4}{5}$, 0.815
$\frac{4}{5}$, 81%, 0.815

10. What is 70% of 80?   56

11. Six is $\dfrac{1}{8}$ of what?   48

12. Six is $\dfrac{2}{5}$ of what?   15

13. $\left(5 - 1\dfrac{2}{3}\right) - 1\dfrac{1}{2} = 1\dfrac{5}{6}$

14. $2\dfrac{2}{5} \div 1\dfrac{1}{2} = 1\dfrac{3}{5}$

15. $0.625 \times 2.4 = 1.5$

16. The prime factorization of 24 is 2 × 2 × 2 × 3, which we can write 2^3 × 3. Write the prime factorization of 36 using exponents. $2^2 \times 3^2$

17. Eight hours and twenty minutes is how many minutes?
500 min.

18. What is the area of this pentagon?
80 sq. in.

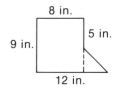

19. Write the standard numeral for 6×10^5. 600,000

20. The diameter of a bike wheel is 20 inches. What is the distance around the wheel? (Use $\pi = 3.14$.) 62.8 in.

21. How many sugar cubes would be needed to build this prism? 60

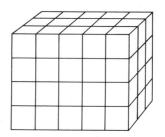

22. The measure of an angle is 60°. Is the angle an acute angle or an obtuse angle? acute

23. Round 3,429,874,102 to the millions' place. 3,430,000,000

24. $(-8) + (+2) = -6$

25. What fraction of the circle is not shaded? $\frac{3}{4}$

LESSON
123

Square Root of Perfect Squares

"What number multiplied by itself equals 25?" We ask this question when we write $\sqrt{25}$. The symbol $\sqrt{}$ is a **square root** sign. We read $\sqrt{25}$ as "the square root of 25." Square root is a geometric idea. The square root of 25 is the length of the side of a square which has an area of 25 square units.

A square with an area of 25 square units has sides which are 5 units long. Therefore, the square root of 25 is 5. Certain numbers are **perfect squares**. We can see by illustration that the numbers 1, 4, 9, and 16 are perfect squares.

We can find the square roots of perfect squares.

$$\sqrt{1} = 1 \qquad \sqrt{4} = 2 \qquad \sqrt{9} = 3 \qquad \sqrt{16} = 4$$

example 123.1 $\quad \sqrt{64} =$

solution The square root of 64 can be thought of in two ways:

1. What is the side of a square which has an area of 64? or
2. What number **multiplied** by itself equals 64?

Either way, we should find the answer is **8**.

practice **a.** The first five perfect squares are 1, 4, 9, 16, and 25. What are the next five perfect squares?   36, 49, 64, 81, 100

b. $\sqrt{81} = 9$ **e.** $\sqrt{100} = 10$

c. $\sqrt{64} = 8$ **f.** $\sqrt{144} = 12$

d. $\sqrt{36} = 6$ **g.** $\sqrt{400} = 20$

problem   **1.** A 32-ounce box of cereal can fill how many 5-ounce bowls?
set 123   6

2. Which time is shortest?
9.99 sec., 10.0 sec., 9.8 sec., 9.85 sec.
9.8 sec.

3. What percent of this circle is **not** shaded?   75%

4. Three is $\dfrac{1}{5}$ of what number?   15

5. Write the standard numeral for $(5 \times 10^6) + (3 \times 10^3)$.
5,003,000

6. Write one hundred and five thousandths as a decimal numeral.   100.005

7. Estimate the product of 496 and 304.   150,000

8. Compare: 55% of 200 ◯ 100   >

Complete the chart to answer problems 9 and 10.

	FRACTION	DECIMAL	PERCENT
		a. 0.1	**b.** 10%
9.	$\dfrac{1}{10}$		
10.	**a.** $1\frac{9}{10}$	1.9	**b.** 190%

11. a. What is 1% of 603?   6.03   **12.** $3\dfrac{1}{2} + 2\dfrac{3}{4} + 5\dfrac{5}{8} = 11\dfrac{7}{8}$

   b. What is 70% of 603?   422.1

13. $\frac{3}{5} \times 1\frac{2}{3} \times 2.5 =$ (fraction answer) $2\frac{1}{2}$

14. $0.2 - (1 - 0.875) = 0.075$ **15.** $0.144 \div (2 \div 0.25) = 0.018$

16. If a 16-ounce can of soup costs 64 cents, what is the cost per ounce?   4 cents

*17. $\sqrt{64} = 8$

18. What is the volume of this prism? 6000 cu. mm

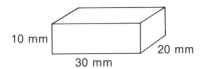

19. What is the circumference of a circle which is 12 inches across? (Use $\pi = 3.14$.) 37.68 in.

20. How many degrees is the measure of an angle which is $\frac{1}{3}$ of a straight angle? 60°

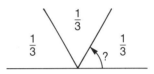

21. $3 \times (4 + 5) = 12 + s$ $s = 15$

22. What is the largest two-digit prime number? 97

23. The number 200 billion is a very large number. Write that number. 200,000,000,000

24. $(4) + (-7) = -3$

25. A function machine does not multiply by 1, but it does the same thing to every number you put into it. Find the rule for this function machine and use it to find the missing number. 21

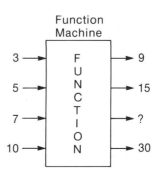

LESSON
124

Writing Units with Exponents

$$\text{ft.} \cdot \text{ft.} = \text{ft.}^2 \qquad \text{m} \cdot \text{m} \cdot \text{m} = \text{m}^3$$

We have seen that units can be reduced (canceled). Units can also be multiplied and combined by using exponents. The area of a rectangle which is 4 inches long and 3 inches wide is 12 square inches.

4 in.

3 in. $\qquad$ 3 in. × 4 in. = 12 sq. in.

We can also use exponents to write square inches because the units are multiplied together.

$$3 \text{ in.} \times 4 \text{ in.} = 3 \times 4 \times \text{in.} \times \text{in.} = 12 \text{ in.}^2$$

All areas are measured with square units and can be written by using exponents as in.^{2}, ft.^{2}, yd.^{2}, cm^2, m^2, km^2, etc.

We can also use exponents to indicate cubic units used to measure volumes. A rectangular prism which is 4 cm × 3 cm × 2 cm has a volume of 24 cubic cm. With the exponents it looks like this.

$$(4 \text{ cm})(3 \text{ cm})(2 \text{ cm}) = 4 \times 3 \times 2 \times \text{cm} \times \text{cm} \times \text{cm} = 24 \text{ cm}^3$$

example 124.1   Each edge of a cube measures 4 ft. Write the volume of the cube with the units in exponent form.

solution   The volume of a cube is found by multiplying the three perpendicular dimensions together.

$$(4 \text{ ft.})(4 \text{ ft.})(4 \text{ ft.}) = 64 \text{ (ft.)(ft.)(ft.)} = \mathbf{64 \text{ ft.}^3}$$

practice Write the units with exponents.

    a. 16 sq. ft. 16 ft.^{2}

    b. 27 cu. m 27 m^3

    c. 24 cm × 10 cm 240 cm^2

    d. 3 m × 4 m × 5 m 60 m^3

    e. The area of a 3 in. × 8 in. rectangle 24 in.^{2}

    f. The volume of a rectangular prism which is 15 cm long, 10 cm wide, and 4 cm high 600 cm^3

problem set 124

1. If 52 cards are dealt out to 7 people as evenly as possible, how many people will end up with 8 cards? 3

2. About how long is a new pencil? 1.8 cm, 18 cm, 180 cm
18 cm

3. What percent of the circles are shaded? 30%

4. $\frac{3}{4} \neq \frac{9}{16}$ (a) $\frac{9}{12}$ (b) $\frac{9}{16}$ (c) $\frac{12}{16}$ (d) $\frac{1\frac{1}{2}}{2}$

5. What is 25% of 48? 12

6. Write the numeral one hundred five thousand, sixty and five hundredths. 105,060.05

7. Another name for the average is the **mean**. What is the mean of 17, 24, 27, and 28? 24

8. Arrange in order from least to greatest: 6.1, $\sqrt{36}$, $6\frac{1}{4}$
$\sqrt{36}$, 6.1, $6\frac{1}{4}$

9. Nine cookies were left in the package. That was $\frac{3}{10}$ of the original number of cookies. How many were in the package originally? 30

10. Twelve is one third of what number? 36

11. Twelve is $\frac{3}{4}$ of what number? 16

12. $2\frac{2}{3} + \left(5\frac{1}{3} - 2\frac{1}{2}\right) = 5\frac{1}{2}$

13. $6\frac{2}{3} \div 4\frac{1}{6} = 1\frac{3}{5}$

14. $4\frac{1}{4} + 3.2 =$ (decimal answer) 7.45

15. $1 - (0.12 \times 1.23) =$ 0.8524

16. How many ounces is 2.5 pounds? (1 pound = 16 ounces.)
40 oz.

17. An angle which measures 100° is what type of angle?
obtuse

18. What is the perimeter of this polygon?   14 cm

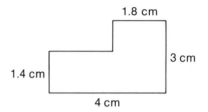

19. A square has a total of how many lines of symmetry?   4

*20. Each edge of a cube measures 4 feet. Write the volume of the cube with units in exponent form.   64 ft.³

21. Complete the proportion: $\dfrac{f}{12} = \dfrac{12}{16}$ 9

22. The function machine is changing the size of these numbers. What number is missing?   8

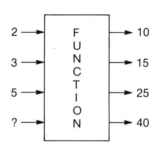

23. Write the standard numeral for 1.25×10^4.   12,500

24. $(-4) + (7) = +3$

25. How many centimeters long is the line segment? (Write the answer as a decimal numeral.)   2.6 cm

LESSON
125

Evaluating Expressions

We have used letters to stand for unknown numbers in problems like $15 + a = 27$. Many times we will see formulas written with all letters or with a combination of numbers and letters, such as $P = 2(l + w)$, and $A = \frac{1}{2}bh$ or $d = rt$. These formulas help us to describe and solve certain common problems.

In this lesson we will practice finding the values of some formulas when we know the values of the letters. Remember that when letters are next to each other with no sign in between, the values are to be multiplied.

example 125.1 What is the value of A in $A = \frac{1}{2}bh$ when b is 8 cm and h is 5 cm?

solution We will rewrite $A = \frac{1}{2}bh$ using numbers in parentheses instead of letters: $A = \frac{1}{2}(8 \text{ cm})(5 \text{ cm})$. Multiplying $\frac{1}{2} \times 8 \times 5$ gives 20 and multiplying cm $\times$ cm gives cm². $A = $ **20 cm²**.

example 125.2 What is the circumference of a circle that has a radius of 5 meters? ($C = 2\pi r$; use $\pi = 3.14$.)

solution $C = 2\pi r$ is a formula for the circumference of a circle. The circumference equals 2 times π times the radius. $C = (2)(3.14)(5 \text{ m}) = $ **31.4 m**.

practice
a. $d = rt$ Find d when r is 45 miles per hour and t is 8 hours.   360 miles

b. $A = s^2$ Find A when s is 9 feet.   81 ft.²

c. $A = \frac{1}{2}bh$ Find A when b is 12 inches and h is 8 inches.   48 in.²   160 mm

d. $P = 2(l + w)$ Find P when l is 30 mm and w is 50 mm.

e. $V = lwh$ Find V when l is 4 m, w is 2 m, and h is 3 m. 24 m³

problem set 125

1. How many trips would it take to fill a 70-gallon water tank with a 4-gallon water bucket?   18 trips

2. What time is 6 hours and 23 minutes before 8:00 P.M.? 1:37 P.M.

3. What percent of the circles are **not** shaded? ·25%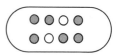

4. Fourteen of the 32 students in the class are girls. What is the boy-girl ratio in the class? $\frac{9}{7}$

5. What is 75% of 16? 12

6. Three out of four is the same as how many out of 100? 75

7. Fifteen is halfway between 7 and what number? 23

8. Compare: $\frac{1}{2}$ of 8 $\bigcirc$ $\frac{1}{4}$ of 16 =

Complete the chart to answer problems 9, 10, and 11.

	FRACTION	DECIMAL	PERCENT
		a.	**b.**
9.	$1\frac{2}{5}$	1.4	140%
10.	**a.** $\frac{1}{10}$	0.1	**b.** 10%
11.	**a.** $1\frac{1}{10}$	**b.** 1.1	110%

12. $3\frac{1}{2} + 2.4 + \frac{3}{5} =$ (fraction answer) $6\frac{1}{2}$

13. $5 \times \frac{3}{4} \times 2\frac{2}{3} = 10$ 14. $1 - (0.1 - 0.01) = 0.91$

15. $0.9 \div (3 \div 0.5) = 0.15$

16. Three and one-half hours is how many minutes? 210 min.

17. $4^2 + \sqrt{4} = 18$

18. What is the area of the shaded portion of this parallelogram? 27 cm²

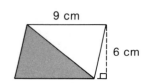

*19. What is the circumference of a circle that has a radius of 5 meters? ($C = 2\pi r$; use $\pi = 3.14$.)   31.4 m

20. For this triangular prism, list the number of its (a) faces, (b) edges, (c) vertexes.   (a) 5,   (b) 9,   (c) 6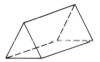

*21. What is the value of A in $A = \frac{1}{2}bh$ when $b = 8$ cm and $h = 5$ cm?   20 cm^2

22. Ten is $\dfrac{2}{3}$ of what number?   15

23. In the figure we show one line of symmetry of an equilateral triangle. An equilateral triangle has a total of how many lines of symmetry?   3

24. $(-12) + (-8) = -20$

25. If one angle of a triangle measures 100°, then what type of a triangle is it?   obtuse

LESSON
126

Opposites and Subtracting Integers

Opposites are numbers which can be written with the same digits but with opposite signs. The opposite of 3 is -3, and the opposite of -5 is 5 (which can be written $+5$).

Opposites are the same distance from zero on the number line, but they lie in opposite directions.

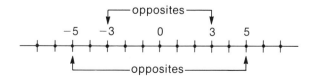

If opposites are added together, the sum is zero.

$$(3) + (-3) = 0 \qquad (-5) + (5) = 0$$

We use opposites to help us subtract integers. To subtract integers we can play the same game we used to add integers, but we need another rule. Instead of subtracting a number, we can add its opposite. The problem $(-5) - (+3)$ can be written $(-5) + (-3)$. We change the subtraction symbol to an addition symbol and reverse the sign of the number being subtracted. The advantage of doing this will be learned later. For our game we will just remember the rule.

> **Instead of subtracting, add the opposite.**

example 126.1   Subtract: $(-4) - (-2) =$

solution   To begin we rewrite the subtraction prob- $\quad (-4) - (-2) =$
lem as an addition problem and reverse $\quad (-4) + (+2) = \mathbf{-2}$
the sign of the number being subtracted.
Now we can play Sign Wars to find out
who survives.

practice   Write the opposite of each.   Subtract.
(See Practice   **a.** -3 3   **b.** 2 -2 **c.** $+5$ -5 **e.** $(-3) - (-4) = +1$
Set RR in the   **f.** $(-4) - (+2) = -6$
Appendix.)   **d.** What is the sum of a number and   **g.** $(+3) - (-6) = 9$
its opposite? 0

problem
set 126

1. If a 36-inch-long sandwich is cut into 8 equal lengths, exactly how long is each length? $4\frac{1}{2}$ in.

2. If photocopies cost 5¢ each for the first hundred copies and 3¢ each for additional copies, what would be the cost of 150 copies? $6.50

3. Thirty people were invited to the party, but only 24 people came. What percent of the people who were invited came to the party? 80%

4. What is 35% of 80? 28

5. Complete the proportion: $\dfrac{7}{20} = \dfrac{w}{100}$ 35

6. Write ten thousand, one hundred fifty-two and six tenths as a decimal number.   10,152.6

7. Divide $10.00 by 7 and round the quotient to the nearest cent. $1.43

8. Arrange in order from least to greatest: $\dfrac{1}{3}$, 32%, 0.3
0.3, 32%, $\frac{1}{3}$

9. Six is $\dfrac{3}{10}$ of what number?   20

10. If $\frac{3}{8}$ of the candy bar weighs 90 grams, then what does the whole candy bar weigh?   240 grams

11. Only 10 people won prizes. This was $\frac{1}{100}$ of those who entered the contest. How many people entered the contest?
1000

12. $12\dfrac{1}{2} - 3\dfrac{5}{8} = 8\dfrac{7}{8}$ **13.** $1\dfrac{2}{3} \div 3\dfrac{1}{2} = \dfrac{10}{21}$

14. $(1 + 0.5) \times (1 - 0.5) =$ 0.75   **15.** $5^2 - \sqrt{25} =$ 20

16. $(-4) + (-2) =$ -6 ***17.** $(-4) - (-2) =$ -2

18. Forty is what fraction of 50? $\frac{4}{5}$

19. What is the area of the pentagon?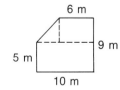
82 m^2

20. The area of a square is 25 m^2. What is the perimeter of the square?   20 m

21. A pyramid with a rectangular base has how many more edges than vertexes?   3

22. Write the numeral for $\frac{1}{4}$ million. 250,000

23. $5y + 5 = 25$ $y = 4$

24. What is the opposite of 4? −4

25. What number will come out of this function machine when a 14 is put into the machine? 18

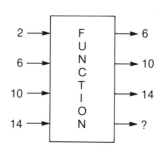

LESSON 127

Finding the Whole When a Decimal Part Is Known

In Lesson 122 we practiced finding a whole when a fraction was known. In this lesson we will practice finding a whole when a decimal part is known.

example 127.1 John counted 12 classmates wearing tennis shoes. If this was 0.4 of the class, how many students were in John's class?

solution If we know part of a whole and what fractional part it is, we can figure out the size of the whole. Remember that a decimal number is a fraction with the denominator shown by the place value of the last digit, which, in this case, is tenths.

The class, then, has been divided into ten parts. Four of the ten parts are wearing tennis shoes. These four parts total 12 students, so each part must equal 3 students. There are ten parts in the whole class, so the whole class is 10 × 3 students = **30 students**.

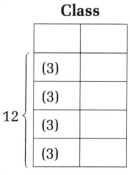

example 127.2 Six is 0.3 of what number?

solution The fraction named is three tenths. The total of the three parts is 6, so each of the three parts must equal 2. Altogether there are ten parts, so the whole is 10 × 2 = **20**.

Whole

$6\begin{cases}(2)\\(2)\\(2)\end{cases}$

practice
(See Practice
Set SS in the
Appendix.)

Draw a diagram for each of these practice problems.
a. Six is 0.2 of what number?   30
b. Eight is 0.1 of what number?   80
c. Eight is 0.4 of what number?   20
d. Ten is 0.2 of what number?   50
e. Forty is 0.8 of what number?   50
f. Three tenths of what number is 12?   40
g. Five tenths of what number is 10?   20
h. Six tenths of what number is 12?   20
i. Nine tenths of what number is 36?   40
j. One hundredth of what number is 5?   500

**problem
set 127**

1. Divide 555 by 12 and write the quotient (a) with a remainder and (b) as a mixed number.   (a) 46 r3, (b) $46\frac{1}{4}$

2. The six gymnasts scored 9.75, 9.8, 9.9, 9.4, 9.9, and 9.95. The lowest score was not counted. What was the sum of the five highest scores?   49.3

3. What time is it 3 hours and 55 minutes after 10:10 A.M.?   2:05 P.M.

4. Eight is $\frac{2}{3}$ of what number?   12

5. Write the standard numeral for $(1 \times 10^5) + (8 \times 10^4) + (6 \times 10^3)$.   186,000

6. Write nine and one half as a decimal number.   9.5

7. The average of three numbers is 12. If two of the numbers are 9 and 10, what is the third number?   17

8. Compare: 26% of 20 $\bigcirc$ 5 $>$

*9. Six is 0.3 of what number? **10.** Ten is 0.2 of what number?
20   50

11. Three tenths of what is 9? **12.** $10^3 - \sqrt{100}$ 990
30

13. $(-15) + (+18) = {}_{+3}$ **14.** $(+8) - (+6) = {}_{+2}$

15. $\left(1\frac{1}{2} + 1\frac{2}{3}\right) \div 3 = {}_{1\frac{1}{18}}$ **16.** $1 \div (1 - 0.975) = {}_{40}$

17. The area of the shaded triangle is 2.8 cm². What is the area of the parallelogram? 5.6 cm²

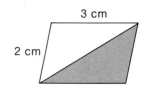

18. What is the volume of a cube which has edges 0.5 meters long? 0.125 m³

19. A rectangle that is not a square has a total of how many lines of symmetry?
2

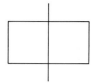

20. If this shape was cut out and folded on the dotted lines, would it form a cube, a pyramid, or a cone? pyramid

21. $5 \cdot (4 + 3) = n + 15$ $n = {}_{20}$

22. What number goes into the function machine to make 32?
8

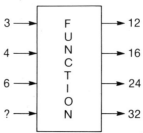

23. How many pounds are in 10 tons? (1 ton = 2000 lb.)
20,000 lb.

24. Which of these polygons is not a quadrilateral?
parallelogram, pentagon, trapezoid, rhombus
pentagon

25. How long is the line
segment?
$1\frac{3}{8}$ in.

LESSON 128

Estimating Square Roots

We are able to find the square roots of numbers which are perfect squares. All of the following numbers are perfect squares: 1, 4, 9, 16, 25, 36, 49, 64, 81, and 100. It is not easy to find the square root of a number that is not a perfect square. For instance, what is the square root of 10? That is, what number multiplied by itself equals 10? The answer is more than 3 because $3 \times 3 = 9$, but it is less than 4 because $4 \times 4 = 16$. We might try $3\frac{1}{2}$ or $3\frac{1}{7}$ or 3.16, but try as we may we cannot find a mixed number or a decimal number which, when multiplied by itself, gives a product exactly equal to 10. In this lesson we will practice finding the two whole numbers between which lie the square roots of numbers which are not perfect squares.

example 128.1   The square root of 10 is between which two whole numbers?

solution   We know that 3×3 equals 9 and that 4×4 equals 16. Thus we know that $\sqrt{10}$ is between **3 and 4**.

$$3 \times 3 = 9$$
$$N \times N = 10$$
$$4 \times 4 = 16$$

practice
(See Practice
Set TT in the
Appendix.)

Each of these square roots is between which two whole numbers?

a. $\sqrt{2}$ 1, 2   **e.** $\sqrt{20}$ 4, 5   **i.** $\sqrt{75}$ 8, 9

b. $\sqrt{5}$ 2, 3   **f.** $\sqrt{30}$ 5, 6   **j.** $\sqrt{90}$ 9, 10

c. $\sqrt{11}$ 3, 4   **g.** $\sqrt{40}$ 6, 7   **k.** $\sqrt{99}$ 9, 10

d. $\sqrt{15}$ 3, 4   **h.** $\sqrt{60}$ 7, 8   **l.** $\sqrt{101}$ 10, 11

problem   **1.** Divide 444 by 16 and write the quotient (a) as a mixed
set 128   number and (b) as a decimal number.
(a) $27\frac{3}{4}$,   (b) 27.75

2. How many years were there from 596 B.C. to 70 B.C.?
526 years

3. Eight is 0.2 of what number?   40

4. $\frac{3}{5} \neq$ (a) 0.6   (b) $\frac{15}{25}$ (c) 60%   (d) 6%   (d) 6%

5. What fraction of 60 is 45? $\frac{3}{4}$

6. Two thirds of an hour is how many minutes?   40 min.

7. Another name for the average is the **mean**. What is the
mean of 12, 10, 9, 9, 8, 7, 7, 7, 6, and 5?   8

8. Compare: 10% of 30 $\bigcirc$ 30% of 10   =

Complete the chart to answer problems 9, 10, and 11.

	FRACTION	DECIMAL	PERCENT
9.	$\frac{12}{25}$	**a.** 0.48	**b.** 48%
10.	**a.** $1\frac{4}{5}$	1.8	**b.** 180%
11.	**a.** $\frac{9}{25}$	**b.** 0.36	36%

12. $3\frac{2}{3} + 1\frac{5}{6} = 5\frac{1}{2}$ **13.** $2\frac{1}{2} \div 1\frac{2}{3} = 1\frac{1}{2}$

14. $0.3 \div (3 \div 0.03) =$ 0.003   **15.** $(-7) + (+16) =$ +9

*16. The square root of 10 is between which two whole
numbers?   3, 4

17. If 2.5 pounds of beef cost $4.50, what is the price per pound? $1.80

18. What is the area of the shaded part of this rectangle? 20 cm²

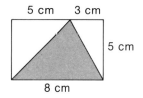

19. What is the name for the shape of a basketball?
sphere

20. $y = 3t + 2$ Find the value of y when t is 2. 8

21. Which of these shapes is *not* symmetrical (has no lines of symmetry)? (c)

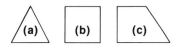

22. Write the prime factorization of 96 using exponents.
$2^5 \cdot 3$

23. Write the standard numeral for 1.25×10^6. 1,250,000

24. $(-6) - (+8) = -14$

25. The graph shows the percentage of the Smiths' income which is spent on various items. The circle represents 100% of their income. What percent of their income goes into savings? 8%

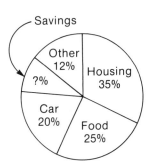

LESSON
129

Unit Conversion

We remember that the name-changer machine changes the name of a number by multiplying the number by a fraction that is equal to 1. This does not change the value. Some fractions that are equal to 1 are:

$$\frac{2}{2}, \qquad \frac{100}{100}, \qquad \frac{100\%}{1}, \qquad \frac{3 \text{ ft.}}{1 \text{ yd.}}, \qquad \frac{100 \text{ cm}}{1 \text{ m}}$$

Whenever the numerator and denominator are equal to each other (and not equal to zero), the fraction is equal to 1. If we know that two numbers are equal, we can use those numbers to write two fractions equal to 1. Since 12 inches equals 1 foot, we can write two fractions that equal 1.

$$\frac{12 \text{ inches}}{1 \text{ foot}} \qquad \text{and} \qquad \frac{1 \text{ foot}}{12 \text{ inches}}$$

Since 5280 feet equal 1 mile, these fractions also equal 1:

$$\frac{5280 \text{ feet}}{1 \text{ mile}} \qquad \text{and} \qquad \frac{1 \text{ mile}}{5280 \text{ feet}}$$

We can change the units of a number by multiplying by a name for 1 which cancels the original units. To change 2.5 miles to yards we can multiply by the fraction which cancels miles and gives an answer in yards.

$$\frac{2.5 \text{ miles}}{1} \times \frac{1760 \text{ yards}}{1 \text{ mile}} = 4400 \text{ yards}$$

example 129.1   Change 4000 ounces to pounds. (1 pound = 16 ounces.)

solution   To change the **name** of this amount, we multiply by a fraction that contains pounds and ounces and that equals 1.

$$\frac{1 \text{ pound}}{16 \text{ ounces}} \quad \text{or} \quad \frac{16 \text{ ounces}}{1 \text{ pound}}$$

In this example we want to cancel ounces, so we multiply by the fraction which has ounces on the bottom.

$$\frac{4000 \text{ ounces}}{1} \times \frac{1 \text{ pound}}{16 \text{ ounces}} = \frac{4000 \text{ pounds}}{16} = \textbf{250 pounds}$$

practice

a. Change 36 feet to yards. (3 feet = 1 yard.)   12 yd.

b. Change 4.2 meters to centimeters. (1 meter = 100 centimeters.)   420 cm

c. Change 364 days to weeks. (7 days = 1 week.)   52 weeks

d. Change 1.2 kilograms to grams. (1 kilogram = 1000 grams.)   1200 grams

e. Change 6000 hours to days. (1 day = 24 hours.)   250 days

f. Change 3 miles to feet. (1 mile = 5280 feet.)   15,840 ft.

problem set 129

1. Fifty-eight players were divided as evenly as possible into 8 teams. How many teams had exactly 7 players?   6

2. To the nearest cent, what is $\frac{1}{3}$ of $10? $3.33

3. What percent of the letters in the word Alaska are A's?   50%

4. Using 1760 yards = 1 mile, write two fractions equal to 1. $\frac{1760 \text{ yd.}}{1 \text{ mi.}}, \frac{1 \text{ mi.}}{1760 \text{ yd.}}$

5. Use digits to write the number fifteen million, nine hundred eighty-five thousand.   15,985,000

*6. Change 4000 ounces to pounds. (1 pound = 16 ounces.)   250 pounds

7. Arrange in order from greatest to least: 5, $\sqrt{26}$, 4.99 $\sqrt{26}$, 5, 4.99

8. What is 6% of 200?   12

9. Fifteen is 0.3 of what number?   50

10. Fifteen is $\frac{5}{6}$ of what number?   18

11. $(8) + (-13) = -5$

12. $(6.2 + 0.16) \times (0.4 - 0.12) =$ 1.7808

13. $4\dfrac{1}{2} \div 0.09 =$ (decimal answer) 50

14. $\dfrac{7}{8} + \left(4 - 1\dfrac{1}{4}\right) = 3\dfrac{5}{8}$ **15.** $2\dfrac{1}{2} \div \left(1\dfrac{1}{2} \times 1\dfrac{2}{3}\right) = 1$

16. Between which two whole numbers is $\sqrt{12}$?   3, 4

17. The 48 people took up only $\frac{2}{3}$ of the bus. How many people can the bus hold?   72

The following scores were made on a test: 72, 80, 84, 88, 100, 88, and 76. Using these scores, answer problems 18, 19, 20, and 21.

18. Which score was earned most often?   88

19. If the scores were listed in order, what would be the middle score?   84

20. What is the average of all the scores?   84

21. What is the difference between the highest score and the lowest score?   28

22. What will come out of this function machine when a ten is put into it?   20

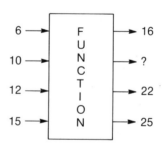

23. $(-8) - (+20) = $ −28   **24.** $(-8) + (-20) = $ −28

25. At the local track the ratio of joggers to walkers was 2 to 1. If eight people were walking, how many were jogging?
16

LESSON
130

Circle Measure—Area

The length of each side of the square on the left is 3 units. The area of the square is 3^2 or 3×3 which equals 9 square units. The next square has sides that are 2 units long. The area of this square is 2^2 which equals 4 square units. The next square has sides that are r long. The area of this square is r^2 square units. On the right the radius of the circle is r units. Each of the four squares has an area of r^2. We see that the combined area of all four of these squares is greater than the area of the circle. The area of three of these squares is less than the area of the circle. **The area of the circle is exactly equal to the area of π of these squares.**

$$\text{Area of circle} = \pi r^2$$

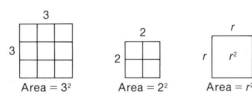 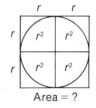

To find the area of a circle, we find the area of a square whose sides equal the radius of the circle. Then we multiply that area by π.

example 130.1 The radius of a circle is 3 cm. What is the area of the circle? (Use $\pi = 3.14$.)

solution We will find the area of a square whose sides equal the radius, then multiply that by 3.14.

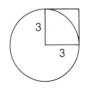

$$\text{Area of square} = 3 \text{ cm} \times 3 \text{ cm} = 9 \text{ cm}^2$$
$$\text{Area of circle} = (3.14)(9 \text{ cm}^2) = \textbf{28.26 cm}^2$$

practice
(See Practice
Set UU in the
Appendix.)

Find the area of the circles with the following measures. (Use $\pi = 3.14$.)

a. Radius, 10 cm   314 cm^2
b. Radius, 2 m   12.56 m^2
c. Radius, 1 km   3.14 km^2
d. Radius, 4 in.   50.24 in.^{2}

e. Diameter, 2 m   3.14 m^2
f. Diameter, 10 yd.   78.5 yd.^{2}
g. Radius, 100 in.   31,400 in.^{2}
h. Diameter, 40 cm   1256 cm^2

**problem
set 130**

1. The quotient of a division problem is 5 and the remainder is 7. The divisor must have been larger than what number?
   7

2. The four judges awarded scores of 9.9, 9.8, 9.6, and 10.0 to the contestant. The highest and lowest scores were not counted. What was the average of the two middle scores?
   9.85

3. What is 32% of 50?   16

4. Using 3 feet = 1 yard, write two ratios equal to 1.
   $\dfrac{1 \text{ yd.}}{3 \text{ ft.}}, \dfrac{3 \text{ ft.}}{1 \text{ yd.}}$

5. What decimal number is three tenths more than twenty-five thousandths?   0.325

6. Change 800 rods to furlongs. (1 furlong = 40 rods.)
   20 furlongs

7. Round the product of $3\frac{1}{3}$ and $2\frac{1}{3}$ to the nearest whole number.   8

8. Compare: $\dfrac{1}{2}$ of 17 $\bigcirc$ 50% of 17   =

9. After reading page 132, Susan figured she had read $\frac{3}{4}$ of her book. How many pages are in her book?   176 pages

10. Twenty is 0.4 of what number?   50

11. Thirty is 0.3 of what number?   100

12. $4\dfrac{1}{2} + 3\dfrac{1}{4} + 2\dfrac{1}{6} = 9\dfrac{11}{12}$

13. $3\dfrac{1}{3} \times 2\dfrac{1}{3} = 7\dfrac{7}{9}$

14. $1.2 + (4 - 1.86) = 3.34$

15. $3.3 \times 2.3 = 7.59$

16. $(-12) + (-8) = -20$

17. $(-3) - (+12) = -15$

*18. The radius of a circle is 3 cm. What is the area of the circle? (Use $\pi = 3.14$.)   28.26 cm²

19. "If two shapes are congruent, then they are similar." True or false?   true

20. $40 + 50 + a = 180$ $a = 90$

21. Find the value of y in $y = mw + 2$ when $m = 3$ and $w = 4$.
14

22. What is the least common multiple (LCM) of 4, 6, and 8?
24

23. The park had a season attendance of 3.5 million people. Write the standard numeral for that number.
3,500,000

24. What number is 5 less than half of 32?   11

25. Which arrow could be pointing to $\sqrt{3}$?   D

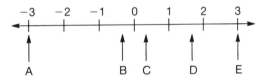

LESSON
131

Probability

Probability is the likelihood that a particular event will happen. For some events the probability can be determined only by an educated guess. For other events we can be more exact.

Probability is written as a ratio. The probability ratio is described as the number of favorable outcomes over the number of possible outcomes.

$$\frac{\text{Number of favorable outcomes}}{\text{Number of possible outcomes}}$$

To help us understand this we will consider two examples.

example 131.1   What is the probability of a coin landing heads up on one toss?

solution   A coin may land either heads up or tails up. So there are 2 **possible** outcomes. The favorable outcome in this question is heads up so there is only 1 **favorable** outcome. The probability is

$$\frac{\text{Favorable}}{\text{Possible}} = \frac{1}{2}$$

example 131.2   What is the probability of drawing an ace from a normal deck of 52 cards? (There are 4 aces in a deck.)

solution   There are 52 **possible** cards one may draw. There are 4 **favorable** cards. The probability is $\frac{4}{52}$, which we can reduce to $\frac{1}{13}$.

practice   From a bag containing only one red, two white, and three blue marbles, what is the probability of drawing:

a. a white marble $\frac{1}{3}$ **d.** a black marble   0

b. a blue marble $\frac{1}{2}$ **e.** a marble   1

c. a red marble $\frac{1}{6}$

problem set 131

1. Compare: $300 \div 17 \bigcirc 300 \div 18$ >

2. If the taxi charges \$1.25 for the first mile and 95¢ for each additional mile, what would be the cost of a 7-mile ride?   \$6.95

3. Change 5000 decimeters to centimeters. (10 centimeters = 1 decimeter.)   50,000 cm

4. Complete the proportion: $\frac{15}{20} = \frac{w}{60}$ $w = 45$

5. Twenty-four is $\frac{2}{3}$ of what number?   36

6. Write the decimal numeral for one hundred and twelve hundredths.   100.12

7. Albert noticed that the cookies could be separated into 2 equal groups, 3 equal groups, or 4 equal groups. If there were less than 20 cookies, how many cookies were there?   12

8. Which of these numbers is closest to 1000?
999, 1001, 999.1, 1001.1   999.1

Complete the chart to answer problems 9, 10, and 11.

	FRACTION	DECIMAL	PERCENT
9.	$\dfrac{3}{100}$	**a.** 0.03	**b.** 3%
10.	**a.** $2\frac{7}{10}$	2.7	**b.** 270%
11.	**a.** $1\frac{9}{20}$	**b.** 1.45	145%

12. $6\dfrac{1}{4} - 1\dfrac{3}{8} = 4\dfrac{7}{8}$ **13.** $5\dfrac{1}{3} \div 2 = 2\dfrac{2}{3}$

14. $5\dfrac{1}{4} + 2.3 =$ (decimal answer)   7.55

15. $0.25 \div (2 \div 0.04) =$ 0.005

16. Arrange in order of size from least to greatest: 2, $\sqrt{3}$, 1^3
$1^3, \sqrt{3}, 2$

17. What is the area of this polygon?
43 ft.^{2}

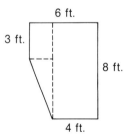

6 ft.

3 ft.

8 ft.

4 ft.

18. What is the area of this circle?
314 m^2

19. What is the circumference of this
circle?   62.8 m

10 m

20. A soup can is what geometric shape?   cylinder

*21. What is the probability of a coin landing heads up on one toss? $\frac{1}{2}$

22. $(-15) + (+24) = $ +9   23. $(-8) - (+3) = $ -11

24. What is the smallest prime number?   2

25. Which of the following shapes could not fold into a cube?

a.

b.

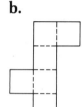

c.

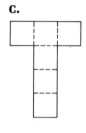

a

LESSON
132

Geometric Formulas

We have practiced finding the perimeters and areas of parallelograms, of special parallelograms called rectangles, of special rectangles called squares, and of triangles. We have studied the formulas for the circumference and the area of a circle. The table below lists formulas for the perimeter and area of each of these shapes.

SHAPE	PERIMETER	AREA
Square	$P = 4s$	$A = s^2$
Rectangle	$P = 2(l + w)$	$A = lw$
Parallelogram	$P = 2(b + s)$	$A = bh$
Triangle	$P = s_1 + s_2 + s_3$	$A = \frac{1}{2}bh$
Circle	$C = \pi d$ or $C = 2\pi r$	$A = \pi r^2$

The letters in the formulas are abbreviations for parts of the shapes as shown below.

A	area	h	height	s	side
b	base	l	length	s_1	side one
C	circumference	P	perimeter	w	width
d	diameter	r	radius	π	pi (≈ 3.14)

practice **a.** What is the formula for the area of a triangle? $A = \frac{1}{2}bh$
b. What is the formula for the perimeter of a square? $P = 4s$
c. The notation $A = \pi r^2$ is a formula for what? area
d. What is the name for the perimeter of a circle? circumference
e. Two congruent triangles can always be arranged to form what geometric figure? parallelogram

problem set 132

1. List the factors of 75. 1, 3, 5, 15, 25, 75

2. About how many millimeters long is your little finger?
0.5, 5, 50, 500 50 mm

3. If 3 parts weigh 24 grams, how much do all 8 parts weigh? 64 grams

24 grams

4. Complete the proportion: $\dfrac{4}{w} = \dfrac{20}{100}$ 20

5. Write the standard numeral for $(7 \times 10^3) + (4 \times 10^0)$.
7004

6. Write the standard numeral for two hundred five million, fifty-six thousand. 205,056,000

7. The average of four numbers is 25. Three of the numbers are 17, 23, and 25. What is the fourth number? 35

8. Compare: 60% of 100 ◯ 31% of 200 <

9. Change 24 drams to scruples. (3 scruples = 1 dram.)
72 scruples

10. Twenty-four guests came to the party. This was $\frac{4}{5}$ of those who were invited. How many guests were invited?   30

11. $1\frac{1}{3} + 3\frac{3}{4} + 1\frac{1}{6} = 6\frac{1}{4}$ **12.** $\frac{5}{6} \times 3 \times 2\frac{2}{3} = 6\frac{2}{3}$

13. $5.62 + 0.8 + 4 =$ 10.42   **14.** $0.08 \div (1 \div 0.4) =$ 0.032

15. $(-2) + (-2) + (-2) =$ -6 **16.** $\sqrt{25} - \sqrt{16} =$ 1

17. At \$1.12 per pound, what is the price per ounce? (1 pound = 16 ounces.)   \$0.07

18. The diameter of a circle is 10 m. What is its area? (Use $\pi = 3.14$.)   78.5 m^2

19. If the area of a square is 36 cm^2, what is its perimeter?
24 cm

20. How many sugar cubes would be needed to make this shape?   24

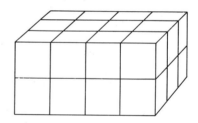

21. Write two formulas for the circumference of a circle.
$C = \pi d$, $C = 2\pi r$

22. What is the probability of guessing the right answer to a multiple choice question if the choices are a, b, c, and d?
$\frac{1}{4}$

23. Write 1.8×10^4 as a standard numeral.   18,000

24. $(-8) - (+7) =$ -15

25. How long is the line segment?
$1\frac{13}{16}$ in.

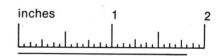

LESSON
133

Finding the Whole When a Percent Is Known

A part of a whole can be written as a fraction, as a decimal fraction, or as a percent. We have worked problems about fractional parts of a whole and about decimal parts of a whole. Now we will consider problems about a percent of a whole. Read the following problem.

Thirty percent of the members of the tribe were warriors. There were 150 warriors in all. What was the population of the tribe?

The problem tells us that 30% of the members of the tribe were warriors. We could work the problem by changing 30% to the fraction $\frac{3}{10}$ or the decimal 0.3. Here we will solve the problem by using percent.

Thirty percent means that the tribe is divided into 100 parts and that 30 of these parts are warriors. Dividing 150 warriors into 30 parts means there are 5 people in each part. The population is all 100 parts. $100 \times 5 = \textbf{500}$

30%

5	5	5	5	5	5	5	5	5	5
5	5	5	5	5	5	5	5	5	5
5	5	5	5	5	5	5	5	5	5
5	5	5	5	5	5	5	5	5	5
5	5	5	5	5	5	5	5	5	5
5	5	5	5	5	5	5	5	5	5
5	5	5	5	5	5	5	5	5	5
5	5	5	5	5	5	5	5	5	5
5	5	5	5	5	5	5	5	5	5
5	5	5	5	5	5	5	5	5	5

100%

practice
(See Practice Set VV in the Appendix.)

Try answering these questions by changing the percent to a fraction or decimal first.

a. Thirty percent of what number is 120?   400
b. Twenty percent of what number is 120?   600
c. Fifty percent of what number is 30?   60
d. Twenty-five percent of what number is 12?   48
e. Twenty is 10% of what number?   200
f. Twelve is 100% of what number?   12
g. Fifteen is 15% of what number?   100

problem set 133

1. Divide 315 by 25 and write the quotient (a) with a remainder, and (b) as a decimal number.
   (a) 12 r15,   (b) 12.6

2. What time is 3 hours and 45 minutes before noon?
   8:15 A.M.

3. If 3 pounds of grapes cost $1.59, what is the cost of 1 pound of grapes? $0.53

4. Use digits to write the number one and one half million.
   1,500,000

5. Round 64,315.28907 to the ten thousandths' place.
   64,315.2891

6. Use digits to write the number one hundred five and five hundredths.   105.05

7. The noontime temperatures for the week were 68°, 70°, 76°, 75°, 76°, 74°, and 72°. What was the mean noontime temperature for the week?   73°

8. Compare: 6% of 50 ◯ 10% of 40   <

9. Fifty percent of what number is 30?   60

10. Thirty percent of the boats in the harbor were capsized by the high winds. If 12 boats were capsized, how many boats were in the harbor?   40

11. What is the probability of getting a six on one roll of a die? (Die is the singular of dice.) $\frac{1}{6}$

12. $5 - \left(4\frac{1}{3} - 1\frac{1}{2} \right) = 2\frac{1}{6}$ **13.** $3\frac{1}{3} \div 2\frac{1}{2} = 1\frac{1}{3}$

14. $3.6 - (0.36 - 0.036) =$
3.276
15. $5.2 \times 3.6 \times 0.27 =$
5.0544

16. $(-4) + (-4) + (-4) = -12$ **17.** $(-15) - (+3) = -18$

18. This isosceles triangle has how many lines of symmetry? 1

19. The square root of 20 is between which two whole numbers? 4, 5

20. Find the value of y if $y = rs + t$ and if $r = 3$, $s = 4$, and $t = 1$. 13

21. What is the greatest common factor (GCF) of 15 and 24?
3

22. Change 400 ounces to pints. (2 pints = 32 ounces.)
25 pints

23. In a bag there are 6 green marbles and 12 red marbles. What is the probability of drawing a green marble from the bag? $\frac{1}{3}$

24. What is the volume of this cube?
216 m^3

25. The volume of the pyramid is $\frac{1}{3}$ of the volume of the cube. What is the volume of the pyramid? 72 m^3

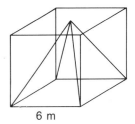

6 m

LESSON

134

Multiplying and Dividing Signed Numbers

The positive and negative numbers are called **signed** numbers. While we are learning the rules for signed numbers we will concentrate on using zero and positive and negative whole numbers. Recall that these numbers are called integers. We can think of signed numbers as having two qualities. One of the qualities is the sign (positive or negative) and the other quality is the numerical part. When we multiply or divide signed numbers, we multiply or divide as usual to find the numerical part of the answer. There are two rules for finding the sign of the answer.

1. If the signs of the two numbers that are multiplied or divided are the same, the answer is positive.
2. If the signs of the two numbers are different, the answer is negative.

examples and solutions

$(+8)(+4) = +32$ $(+8) \div (+4) = +2$
$(+8)(-4) = -32$ $(+8) \div (-4) = -2$
$(-8)(+4) = -32$ $(-8) \div (+4) = -2$
$(-8)(-4) = +32$ $(-8) \div (-4) = +2$

practice
(See Practice Set WW in the Appendix.)

a. $(-3)(2) = -6$ **g.** $(-3)(-4) = +12$ **m.** $(+30) \div (-6) = -5$

b. $(-6)(-12) = +72$ **h.** $(-8) \div (2) = -4$ **n.** $(-30) \div (5) = -6$

c. $(+7)(+6) = +42$ **i.** $(-8) \div (-2) = +4$ **o.** $\dfrac{-36}{9} = -4$

d. $(-8)(+3) = -24$ **j.** $(12) \div (-3) = -4$ **p.** $\dfrac{36}{-6} = -6$

e. $(2)(-5) = -10$ **k.** $(-12) \div (4) = -3$ **q.** $\dfrac{-36}{-12} = +3$

f. $(6)(+7) = +42$ **l.** $(-12) \div (-6) = +2$ **r.** $\dfrac{+100}{-10} = -10$

problem set 134

1. Divide 315 by 20 and write the quotient (a) with a remainder and (b) as a decimal number. (a) 15 r15, (b) 15.75

2. Jenny ran 4 laps in 5 minutes. How many seconds did it take to run each lap if she ran at a steady pace? 75 sec.

3. Fifteen is 25% of what number? 60

4. Complete the proportion: $\dfrac{f}{12} = \dfrac{6}{9}$ $f = 8$

5. Use digits to write the number fifty and forty-nine thousandths. 50.049

6. Compare: $(+6) + (-3) \bigcirc (+6) - (-3)$ $<$

7. Divide 0.23 by 0.7 and round the quotient to the nearest thousandth. 0.329

8. Six out of every 100 fans cheered for the visiting team. If 60 people cheered for the visiting team, how many fans were there? 1000

Complete the chart to answer problems 9, 10, and 11.

	FRACTION	DECIMAL	PERCENT
9.	$\dfrac{3}{6}$	**a.** 0.5	**b.** 50%
10.	**a.** $2\frac{3}{5}$	2.6	**b.** 260%
11.	**a.** $\frac{4}{25}$	**b.** 0.16	16%

12. $6.4 + 3\dfrac{1}{2} =$ (fraction answer) $9\frac{9}{10}$

13. $\dfrac{9}{10} \times \dfrac{5}{12} \times \dfrac{8}{15} = \dfrac{1}{5}$

14. $5.35 + 6 + 2\dfrac{1}{8} =$ (decimal answer) 13.475

15. $14 \div 1.2 =$ (Round to hundredths.) 11.67

16. On the number line $\sqrt{31}$ is between which two whole numbers? 5, 6

At right is a list of scores earned in a diving competition. Use this list to answer problems 17, 18, 19, and 20.

6.0
6.5
7.0
7.5
6.5
6.5
7.0

17. What score was made most often?
6.5

18. If the scores were arranged in order of size, which would be the middle score?   6.5

19. What is the average of all the scores? (Round to the nearest tenth.)   6.7

20. What is the difference between the highest and the lowest score?   1.5

21. Write the prime factorization of 210. $2 \cdot 3 \cdot 5 \cdot 7$

22. What is the probability of getting a seven on one roll of a die?   0

23. It was estimated that three quarters of a billion people watched the closing ceremonies of the Olympics. Write that numeral.   750,000,000

24. (a) $(-6)(+3) =$ (b) $(-6) \div (+3) = -2$
    -18

25. Change 4.2 liters to milliliters. (1 liter = 1000 milliliters.)
    4200 ml

LESSON
135

Divisibility

Divisibility is the "ability" of a number to be divided by another number without a remainder.

The number 159,840 is divisible by which of these numbers?
2, 3, 4, 5, 6, 8, 9, 10

The tests for divisibility help us answer this question without needing to do the actual division. Below are some tests for divisibility. Which of these tests does the number 159,840 pass?

Tests for Divisibility

A number is able to be divided by ...
2   if the last digit can be divided by 2.   4   if the last 2 digits can be divided by 4.   8   if the last 3 digits can be divided by 8.
5   if the last digit is 0 or 5.   10   if the last digit is 0.
3   if the *sum of the digits* can be divided by 3.   9   if the *sum of the digits* can be divided by 9.   6   if the number can be divided by 2 *and* by 3.

We find that 159,840 is divisible by 2, 3, 4, 5, 6, 8, 9, and 10. Thus each of these numbers is a divisor of 159,840.

example 135.1   Which positive one-digit numbers are divisors of 9060?

solution   The positive one-digit numbers are 1, 2, 3, 4, 5, 6, 7, 8, and 9. Using the tests from the chart above, we find that the divisors of 9060 are **1**, **2**, **3**, **4**, **5**, and **6**.

practice   Which positive one-digit numbers are divisors of the following?

a. 180   1, 2, 3, 4, 5, 6, 9   **c.** 3636   1, 2, 3, 4, 6, 9   **e.** 123,456   1, 2, 3, 4, 6, 8

b. 280   1, 2, 4, 5, 7, 8   **d.** 11,115   1, 3, 5, 9   **f.** 101,010   1, 2, 3, 5, 6, 7

problem set 135

1. Divide 315 by 24 and write the quotient (a) as a mixed number and (b) as a decimal number.
(a) $13\frac{1}{8}$,   (b) 13.125

*2. Which positive one-digit numbers are divisors of 9060?
1, 2, 3, 4, 5, 6

3. Six of the 14 players were left-handed. What was the ratio of left-handed players to right-handed players? $\frac{3}{4}$

4. 24% ≠   (a) $\frac{12}{50}$ (b) 0.24   (c) $\frac{6}{25}$ (d) 24   (d) 24

5. Write the standard numeral for $(6 \times 10^7) + (5 \times 10^4)$.
60,050,000

6. Estimate the sum of 6.9, $3\frac{1}{3}$, 7.01, and $3\frac{5}{6}$ to the nearest whole number.   21

7. Cancel units and multiply: 384 ounces

$$\frac{3 \text{ gal.}}{1} \times \frac{4 \text{ qt.}}{1 \text{ gal.}} \times \frac{2 \text{ pt.}}{1 \text{ qt.}} \times \frac{16 \text{ ounces}}{1 \text{ pt.}}$$

8. Arrange in order from greatest to least: $1\frac{1}{2}$, 125%, 1.39
$1\frac{1}{2}$, 1.39, 125%

9. Ten percent of the books were damaged. If 20 books were damaged, how many books were there in all?   200

10. Eight is 20% of what number?   40

11. Eight is $\frac{2}{5}$ of what number?   20

12. What is the probability that the spinner will stop on 3? $\frac{1}{4}$

13. $5\frac{1}{2} - \left(4\frac{1}{4} - 3\frac{1}{3}\right) = 4\frac{7}{12}$ **14.** $3 \div \left(1\frac{2}{3} \div 3\right) = 5\frac{2}{5}$

15. $0.1 - 0.0986 = 0.0014$ **16.** $3.5 \times 1.2 \times 100 = 420$

17. What is the cost per doughnut if two dozen doughnuts cost $4.32? $0.18

18. What is the area of this circle?
1256 m^2

(Continue to use $\pi = 3.14$)

19. What is the circumference of this circle?   125.6 m

20. Each face of a cube is a square. What is the area of each square face of this cube?   25 cm^2

5 cm

21. What is the value of y if $y = t^2 + c$ and if $t = 6$ and $c = 7$?
43

22. Which positive one-digit numbers are **not** divisors of 420?
8, 9

23. If the probability that it will rain is $\frac{1}{4}$, then what is the probability that it will not rain? $\frac{3}{4}$

24. (a) $(-7)(+2) =$ (b) $(-12) \div (-3) =$
  -14 $+4$

25. Which of these numbers is divisible by 2, 3, and 5?
(a) 54,321   (b) 12,345   (c) 543,210   (d) 55,550
(c) 543,210

LESSON
136

Surface Area of a Prism

We have measured the volumes of rectangular prisms and special rectangular prisms called cubes. Besides measuring the volume of a solid we can also measure the total area of all of its surfaces. This is called the **surface area** of a solid. A cube puzzle can help us understand this idea.

How many square stickers are needed to cover this cube? We could just begin counting and hope that we count them all, or we can be systematic in our approach to the problem. How many squares are needed to cover one face? How many faces are there in all? What then is the total number of squares needed to cover all the faces?

example 136.1   How many squares, 1 inch on a side, would be needed to cover this prism?

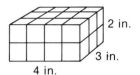

2 in.
3 in.
4 in.

solution   The rectangular prism has 6 surfaces. We must find the area of each surface, then add to find the total surface area. The front and back faces are 2-inch by 4-inch rectangles. The area of each is 8 in.². The top and bottom faces are 3-inch by 4-inch rectangles with areas of 12 in.². The left and right rectangles each have an area of 6 in.². The total surface area is 8 in.² + 8 in.² + 12 in.² + 12 in.² + 6 in.² + 6 in.² = **52 in.²**.

practice   Find the surface area of each rectangular prism.

a.   96 cm²

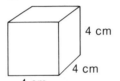

4 cm
4 cm
4 cm

b.   78 ft.²

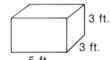

3 ft.
3 ft.
5 ft.

c.   6 m²

1 m
1 m
1 m

problem set 136

1. Divide 938 by 40 and write the quotient (a) as a mixed number and (b) as a decimal number.
   (a) $23\frac{9}{20}$,   (b) 23.45

2. One fifth of the 300 troops were injured. How many were not injured?   240

3. Complete the proportion: $\dfrac{12}{20} = \dfrac{18}{g}$ g = 30

4. Write the standard numeral for twenty million, one hundred thousand, fifty.   20,100,050

5. Twenty is 25% of what number?   80

6. If the mean of 10 numbers is 6.4, what is their sum?
   64

7. What is 1% of one million?   10,000

8. Change 16,000 acres to square miles. (1 square mile = 640 acres.) 25 sq. mi.

9. Write $\dfrac{1}{6}$ as a percent. (End the percent with a fraction.)
$16\frac{2}{3}\%$

10. $(1.2)^2 - 1.2 =$ 0.24

11. What is the probability of drawing a red marble with a single draw from a bag containing one red, two white, and four blue marbles? $\frac{1}{7}$

12. What is the total surface area of a cube if the area of each face is 25 cm²? 150 cm²

13. $\left(3\dfrac{3}{4} + 2\dfrac{1}{2}\right) \times 1\dfrac{1}{3} = 8\frac{1}{3}$ **14.** $\left(3\dfrac{1}{2} - 1\dfrac{2}{3}\right) \div 1\dfrac{1}{3} = 1\frac{3}{8}$

15. (a) $(+12)(-8) =$ (b) $\dfrac{(+12)}{(-6)} = -2$
-96

16. The ratio of orangutans to chimpanzees was 2 to 5. If there were 40 chimpanzees, how many orangutans were there?
16

17. What is the circumference of this circle? 125.6 mm

40 mm

18. What is the area of this circle?
1256 mm²

19. What is the area of this parallelogram? 17.92 in.²

2.8 in.

6.4 in.

20. An octagon has how many more sides than a quadrilateral?
4

21. How many square stickers, 1 inch on a side, would be needed to cover this prism?   90

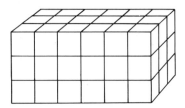

22. Twenty-three thousand is **not** divisible by which of these numbers? 1, 2, 3, 5, 10   3

23. What is the probability of an event that is **certain** to happen?   1

24. $75 + 50 + a = 180$ $a = 55$

25. To which fraction is the arrow pointing? $\frac{1}{3}$

LESSON 137

Sum of the Angles of a Triangle

If we cut circular corners off of any triangle we can fit the corners together to make half a circle, as we show here. This tells us something about the sum of the angles of a triangle. A full circle measures 360°, so a half circle measures 180°. Together the three angles of a triangle equal a half circle, so the sum of their measures is 180°. This is a very important piece of knowledge to remember!

> **The sum of the angle measures of any triangle is 180°.**

example 137.1 The measure of two of the angles of the triangle are given. What is the measure of the third angle?

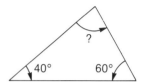

solution The sum of the three angles must total 180°. The two angles given total 100°. Subtracting 100° from 180°, we find that the third angle must measure **80°**.

practice Find the measure of the angle not given.

a. 90°

b. 58°

c. 41°

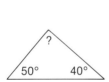
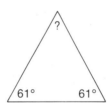
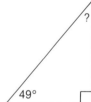

d. If the three angles of a triangle are equal to each other, then each angle measures how many degrees? 60°

problem set 137

1. The league is forming soccer teams so that there are 14 or 15 players on each team. If 300 players sign up for soccer, what is the greatest number of teams which can be formed? 21

2. If the sales tax rate is 6 cents for every one dollar of sales, how much is the tax on a television set which sells for $350? $21

3. What is the area of this circle? 113.04 m²

4. What is the circumference of this circle? 37.68 m

5. Write a quarter of a million as a standard numeral.
250,000

6. If the sum of 12 numbers is 288, what is the average of the numbers?   24

7. How many millimeters is 1.25 meters? (1 meter = 1000 millimeters.)   1250 mm

8. $(1.1)^3 - 1.1 =$ 0.231

Complete the chart to answer problems 9, 10, and 11.

	FRACTION	DECIMAL	PERCENT
9.	$\dfrac{19}{20}$	**a.** 0.95	**b.** 95%
10.	**a.** $1\frac{9}{10}$	1.9	**b.** 190%
11.	**a.** $\frac{16}{25}$	**b.** 0.64	64%

12. $\dfrac{4.32}{0.003} =$ 1440

13. $1\dfrac{2}{3} \div \left(3 \div \dfrac{3}{5} \right) = \dfrac{1}{3}$

14. $0.375 \times 0.16 =$ 0.06

15. $60° + 70° + n = 180°$ $n =$ 50°

16. Compare: $\dfrac{1}{100}$ of one million $\bigcirc$ 1% of one million   =

17. What number is 10 less than the product of 9.6 and 2.04?
9.584

18. What is the probability that the spinner will stop on a number less than 4?   1

19. What is the probability of an event that **cannot** happen?
0

20. What is the measure of angle A?
40°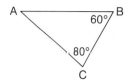

21. How many square tiles 1 foot on a side would be needed to cover this cube? 96

4 ft.

22. What is the volume of this cube?
64 ft.^{3}

The table shows the number of students who received certain scores on a class quiz. The frequency is the number of students who made that score.

23. Which score was made most often? 8

24. How many students scored lower than 8? 13

25. What fraction of the class scored 8 or more? $\frac{12}{25}$

SCORE	FREQUENCY
10	2
9	3
8	7
7	5
6	5
5	3

LESSON
138

Roman Numerals

The numerals we normally use to write numbers are called Arabic numerals. There are other ways to write numbers. One ancient form of numerals which we still use today is Roman numerals. We see Roman numerals used to number chapters in books, to mark hours on clocks, to date buildings, and to number Olympiads and Super Bowl games.

The table below lists values of some Roman numerals.

NUMERAL	I	V	X	L	C	D	M
VALUE	1	5	10	50	100	500	1000

The Roman numeral system does not use place value. The numeral II does not stand for eleven, it stands for two. The value of the numerals are added together unless a numeral of lesser value is written in front of a numeral of greater value, in which case the smaller is subtracted from the larger.

examples and solutions

III = 3   IX = 9   XXX = 30   DC = 600
IV = 4   XIII = 13   XL = 40   MM = 2000
VI = 6   XIV = 14   LXX = 70   MCM = 1900
VIII = 8   XVI = 16   XC = 90   MCMLXXX = 1980

practice
(See Practice Set XX in the Appendix.)

Write the Arabic numeral for:

a. XXIII 23
b. XLII 42
c. CXC 190
d. LXXIX 79
e. CCC 300
f. DCC 700
g. LXXIX 79
h. DCCLXVI 766
i. MDCLXVI 1666
j. MDCCLXXVI 1776

Write the Roman numeral for:

k. 7 VII
l. 19 XIX
m. 24 XXIV
n. 80 LXXX
o. 88 LXXXVIII
p. 140 CXL
q. 400 CD
r. 750 DCCL
s. 900 CM
t. 2001 MMI

problem set 138

1. Divide 596 by 32 and write the quotient as a decimal number rounded to the nearest tenth. 18.6

2. Twenty-seven is $\frac{3}{4}$ of what number? 36

3. Twenty is what percent of 25? 80%

4. Compare: $(0.1)^2 \bigcirc (0.1)^3$ >

5. Write as a decimal numeral nine thousand, one hundred fifty and twenty-five thousandths.   9150.025

Complete the chart to answer problems 6 through 11.

ROMAN NUMERAL	ARABIC NUMERAL	
CCXXXIV	**6.**	234
CLXVI	**7.**	166
MDCXL	**8.**	1640
9. CXXIV	124	
10. DLV	555	
11. MCCXXXIV	1234	

12. $6\dfrac{1}{4} - \left(1\dfrac{7}{8} + 3\dfrac{1}{2}\right) = \dfrac{7}{8}$ **13.** $2\dfrac{1}{2} \div \left(1\dfrac{2}{3} \times 3\right) = \dfrac{1}{2}$

14. $12 \div (0.18 \div 12) = 800$ **15.** $(-6) - (+15) = -21$

16. $(-6)(+15) = -90$

17. Write the prime factorization of 64 using exponents.
2^6

18. What is the area of this pentagon? 195 mm^2

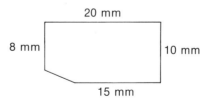

20 mm

8 mm   10 mm

15 mm

19. What is the circumference of this circle?   12.56 ft.

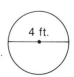

4 ft.

20. What is the area of this circle? 12.56 ft.^2

21. What is the volume of this prism?
24 m³

22. What is the surface area of this prism?
52 m²

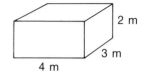

23. The measure of ∠A is 80°, and the measure of ∠B is 50°.
What is the measure of ∠C?   50°

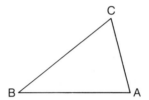

24. What number is 5 more than half the product of 12 and 16?
101

25. To what number is the arrow pointing?   225

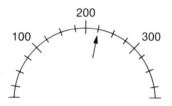

LESSON
139

Angle Measure—
Using a Protractor

In this lesson we will practice using a tool which helps us measure the size of angles. The tool is a **protractor**. To measure an angle we place the center point of the protractor on the vertex and the base line on one side of the angle. Where the other side of the angle passes through the scale we can read the size of the angle.

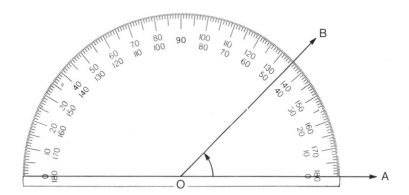

The scale on a protractor has two sets of numbers. One set is for measuring angles starting from the right side, and the other for measuring angles starting from the left. The easiest way to be sure we are reading from the correct scale is to decide if the angle we are measuring is acute or obtuse. Looking at ∠AOB we read the numbers 45° and 135°. Since the angle is less than 90° (acute), it must be 45° and not 135°.

practice
(See Practice
Set YY in the
Appendix.)

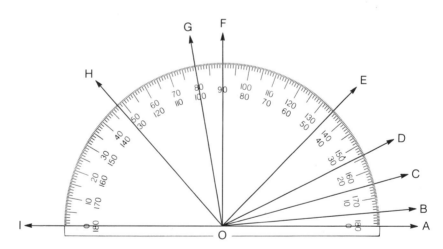

What is the measure of each angle?
a. ∠AOC ₁₅°   **d.** ∠AOG ₁₀₀°   **g.** ∠IOG ₈₀°   **i.** ∠IOB ₁₇₅°
b. ∠AOD ₂₆°   **e.** ∠AOH ₁₃₂°   **h.** ∠IOF ₉₀°   **j.** ∠IOA ₁₈₀°
c. ∠AOE ₄₅°   **f.** ∠IOH ₄₈°

**problem
set 139**

1. Divide 495 by 12 and write the quotient (a) as a mixed number and (b) as a decimal number.
(a) $41\frac{1}{4}$, (b) 41.25

2. In 15 games the team had 9 wins and no ties. What was the team's won–lost ratio? $\frac{3}{2}$

3. Complete the proportion: $\dfrac{15}{25} = \dfrac{x}{100}$ 60

4. Fifteen is $\dfrac{5}{8}$ of what number?   24

5. Fifteen is what percent of 25?   60%

6. What number is 60% of 25?   15

7. Which of these numbers is closest to the average of the numbers?   5, 6, 6, 6, 7, 9, 9   7

Complete the chart to answer problems 8 through 11.

ROMAN NUMERAL	ARABIC NUMERAL
CXLIII	**8.** 143
MCMXXIX	**9.** 1929
10. MCMIX	1909
11. DCCLXIV	764

12. $5\dfrac{2}{3} + 9 + 6\dfrac{5}{6} = 21\dfrac{1}{2}$ **13.** $6\dfrac{2}{3} \times \left(5 \div 1\dfrac{2}{3}\right) = 20$

14. $5\dfrac{3}{4} - 2.6 =$ (decimal answer)   3.15

15. $0.15 \div (6 \div 0.15) = 0.00375$

16. (a) $(-30) + (+6) =$ (b) $(-30) - (+6) =$
     -24 -36

17. (a) $(-30) \times (+6) =$ (b) $(-30) \div (+6) =$
     -180 -5

18. Which two triangles appear to be congruent?
     a, e

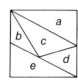

19. Which of these numbers is a composite number?
21, 31, 41, 61   21

Using the protractor, find the measure of each angle.

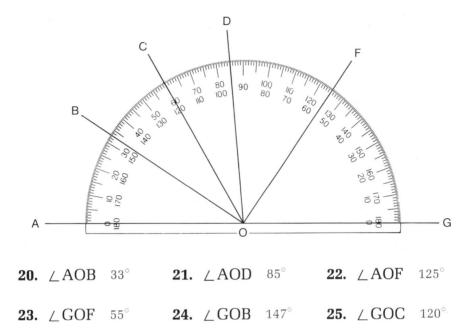

20. ∠AOB   33°   **21.** ∠AOD   85°   **22.** ∠AOF   125°

23. ∠GOF   55°   **24.** ∠GOB   147°   **25.** ∠GOC   120°

LESSON
140

Rectangular Coordinates

By drawing two lines perpendicular to each other and by extending the unit marks, we can create a grid or graph. We can name the location of any point on this graph with two numbers.

The point at which the number lines cross is called the **origin**. The horizontal number line is called the **x-axis**, and the vertical number line is called the **y-axis**. We **graph** a point when we make a dot at the location of the point. The numbers that tell the location of the point are called the **coordinates** of the point. The coordinates are written as a pair of numbers in parentheses, like (3, −2). The first number shows the horizontal (↔) direction and distance from the origin. The second number shows the vertical (↕) direction and distance from the origin. The sign of the coordinate shows direction. Positive co-ordinates are to the right or up, and negative coordinates are to the left or down. We begin at the origin, which has the co-ordinates (0, 0).

example 140.1   Graph the point (3, −2).

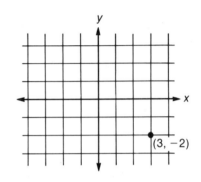

solution   Beginning at the origin we move horizontally positive 3, that is, 3 units to the right. **From there** we move negative 2, which is down 2 units. We draw a dot at that location and label it (3, −2).

practice
(See Practice
Set ZZ in the
Appendix.)

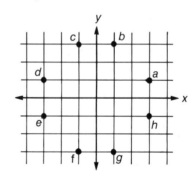

What point has the coordinates:
a. (3, −1)   *h*   **b.** (1, 3)   *b*   **c.** (−3, 1)   *d*
What are the coordinates of:
d. *d*   (−3, 1)   **e.** *e*   (−3, −1) **f.** *f*   (−1, −3)

**problem
set 140**

1. When the product of 0.6 and 0.12 is divided by the sum of 4.98 and 0.02, what is the quotient?   0.0144

2. Thirty-six is what percent of 50?   72%

3. What is the measure of angle C?   75°

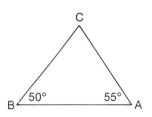

4. What is the area of this circle?
7850 mm²

5. What is the circumference of this circle?
314 mm

6. What is the volume of this cube?
1000 cm³

7. What is the surface area of this cube?
600 cm²

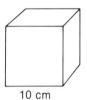

10 cm

Complete the chart to answer problems 8 through 11.

ROMAN NUMERAL	ARABIC NUMERAL
DCCLXV	**8.** 765
MCMXLVII	**9.** 1947
10. DCCCLXXXVIII	888
11. MCMXC	1990

12. $7.6 + 3\frac{1}{2} + 1\frac{1}{10} =$ (fraction answer) $12\frac{1}{5}$

13. $2\frac{1}{3} \div \left(4 - 1\frac{2}{3}\right) = 1$

14. (a) $(-18) + (-6) =$ (b) $(-18) - (-6) =$
-24 -12

15. (a) $(-18)(-6) = +108$ (b) $\dfrac{-18}{-6} = +3$

Use this protractor to find the measure of the angles in questions 16, 17, 18, and 19.

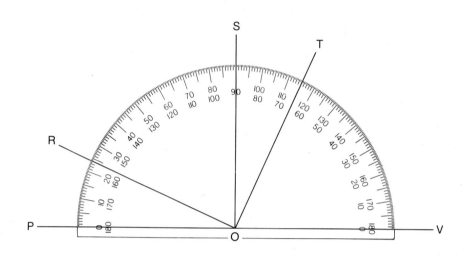

16. ∠POR   24°

17. ∠POS   90°

18. ∠VOS   90°

19. ∠VOT   65°

Which points in the graph below have these coordinates?

20. (−4, 2)   d   **21.** (0, 2)   j   **22.** (2, −4)   g

What are the coordinates of these points?

23. k   (−2, 0)

24. e   (−4, −2)

25. b   (2, 4)

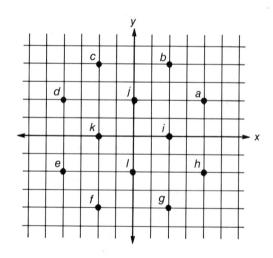

Appendix

Practice Sets

Calculator Exercises

This appendix contains additional practice problems for concepts presented in selected lessons. It is very important that no problems in the regular problem sets be omitted to make room for these problems. This book is designed to produce long-term retention of concepts, and long-term practice of all the concepts is necessary. The practice problems in the problem sets provide enough initial exposure to concepts for most students. If a student continues to have difficulty with certain concepts, some of these problems can be assigned as remedial exercises.

Set A

1. $576 \times 8 =$ 4608
2. $308 \times 7 =$ 2156
3. $784 \times 6 =$ 4704
4. $4306 \times 9 =$ 38,754
5. $42 \times 30 =$ 1260
6. $56 \times 40 =$ 2240
7. $60 \times 78 =$ 4680
8. $70 \times 64 =$ 4480
9. $90 \times 70 =$ 6300
10. $300 \times 60 =$ 18,000
11. $400 \times 50 =$ 20,000
12. $80 \times 500 =$ 40,000
13. $37 \times 43 =$ 1591
14. $62 \times 74 =$ 4588
15. $86 \times 27 =$ 2322
16. $94 \times 63 =$ 5922
17. $408 \times 24 =$ 9792
18. $507 \times 37 =$ 18,759
19. $62 \times 409 =$ 25,358
20. $84 \times 306 =$ 25,704
21. $520 \times 36 =$ 18,720
22. $940 \times 42 =$ 39,480
23. $790 \times 86 =$ 67,940
24. $243 \times 67 =$ 16,281
25. $896 \times 75 =$ 67,200

Set B

Write uneven division with a remainder.

1. $375 \div 5 =$ 75
2. $244 \div 6 =$ 40 r4
3. $892 \div 7 =$ 127 r3
4. $143 \div 4 =$ 35 r3
5. $412 \div 4 =$ 103
6. $423 \div 6 =$ 70 r3
7. $728 \div 7 =$ 104
8. $812 \div 9 =$ 90 r2
9. $1206 \div 6 =$ 201
10. $824 \div 4 =$ 206
11. $906 \div 9 =$ 100 r6
12. $3492 \div 7 =$ 498 r6
13. $144 \div 12 =$ 12
14. $472 \div 10 =$ 47 r2
15. $893 \div 15 =$ 59 r8
16. $762 \div 25 =$ 30 r12
17. $432 \div 20 =$ 21 r12
18. $986 \div 50 =$ 19 r36
19. $1427 \div 25 =$ 57 r2
20. $3819 \div 19 =$ 201
21. $4126 \div 32 =$ 128 r30
22. $968 \div 24 =$ 40 r8
23. $377 \div 18 =$ 20 r17
24. $566 \div 42 =$ 13 r20
25. $1349 \div 37 =$ 36 r17

Set C

Write the numeral named.

1. Five thousand   5000
2. Two hundred eight   208
3. One thousand, two hundred   1200
4. Six thousand, fifty   6050
5. Nine hundred forty-three   943
6. Eight thousand, one hundred ten   8110
7. Ten thousand   10,000
8. Twenty-one thousand   21,000
9. Forty thousand, nine hundred   40,900
10. One thousand, ten   1010

11. Fifteen thousand, twenty-one   15,021
12. Nineteen thousand, eight hundred   19,800
13. One hundred thousand   100,000
14. Two hundred ten thousand   210,000
15. Four hundred five thousand   405,000
16. Three hundred twenty-five thousand   325,000
17. One million   1,000,000
18. One million, two hundred thousand   1,200,000
19. Ten million, one hundred fifty thousand   10,150,000
20. Five hundred million   500,000,000
21. Two million, fifty thousand   2,050,000
22. Twenty-five million, seven hundred fifty thousand   25,750,000
23. Five billion   5,000,000,000
24. One billion, two hundred fifty million   1,250,000,000
25. Twenty-one billion, five hundred ten million   21,510,000,000

Set D

Round each to the nearest ten.
1. 678   680   **2.** 83   80   **3.** 575   580   **4.** 909   910
5. 99   100   **6.** 1492   1490   **7.** 104   100   **8.** 1321   1320

Round each to the nearest hundred.
9. 678   700   **10.** 437   400   **11.** 846   800   **12.** 1587   1600
13. 1023   1000 **14.** 987   1000 **15.** 3679   3700 **16.** 4981   5000

Round each to the nearest thousand.
17. 1986   2000   **18.** 2317   2000   **19.** 1484   1000
20. 3675   4000   **21.** 5280   5000   **22.** 1760   2000
23. 36,102   36,000 **24.** 57,843   58,000 **25.** 375,874   376,000

Set E

Find the average of each set of numbers.
1. 15, 18, 21   18   **2.** 16, 18, 20, 22   19
3. 5, 6, 7, 8, 9   7   **4.** 2, 4, 6, 8, 10, 12   7

5. 100, 200, 300, 400   250   **6.** 20, 30, 40, 50, 60   40
7. 23, 35, 32   30   **8.** 136, 140, 141   139
9. 94, 94, 98, 98   96   **10.** 68, 72, 68, 76, 76   72
11. 6847, 6951   6899   **12.** 86, 86, 86, 86, 86   86
13. 562, 437, 381   460   **14.** 6, 6, 3, 9, 3, 6, 9   6
15. 34, 41, 46, 50, 54   45

What number is halfway between each pair of numbers?
16. 47, 91   69   **17.** 56, 88   72   **18.** 75, 57   66   **19.** 1, 101   51
20. 92, 136   114   **21.** 253, 325   289   **22.** 548, 752   650   **23.** 1776, 1986   1881
24. 101, 909   505   **25.** 432, 234   333

Set F

List the factors of each number.
1. 30   1, 2, 3, 5, 6, 10, 15, 30   **2.** 40   1, 2, 4, 5, 8, 10, 20, 40   **3.** 50   1, 2, 5, 10, 25, 50
4. 60   1, 2, 3, 4, 5, 6, 10, 12, 15, 20, 30, 60   **5.** 35   1, 5, 7, 35   **6.** 36   1, 2, 3, 4, 6, 9, 12, 18, 36
7. 37   1, 37   **8.** 38   1, 2, 19, 38   **9.** 39   1, 3, 13, 39   **10.** 49   1, 7, 49

Find the Greatest Common Factor (GCF) of each set of numbers.
11. 14, 28   14   **12.** 12, 20   4   **13.** 15, 16   1
14. 15, 25   5   **15.** 25, 50   25   **16.** 40, 70   10
17. 24, 42   6   **18.** 12, 21   3   **19.** 22, 55   11
20. 12, 30   6   **21.** 4, 3, 2   1   **22.** 2, 4, 6   2
23. 4, 8, 12   4   **24.** 6, 12, 20   2   **25.** 24, 30, 12, 15   3

Set G

1. $\frac{1}{2}$ of 42 = 21   **2.** $\frac{1}{3}$ of 42 = 14

3. $\frac{2}{3}$ of 42 = 28   **4.** $\frac{1}{4}$ of 60 = 15

5. $\frac{3}{4}$ of 60 = 45   **6.** $\frac{2}{3}$ of 60 = 40

7. $\dfrac{1}{5} \times 60 =$ 12   **8.** $\dfrac{2}{5} \times 60 =$ 24

9. $\dfrac{3}{5} \times 20 =$ 12   **10.** $\dfrac{3}{8} \times 24 =$ 9

11. $\dfrac{5}{6} \times 24 =$ 20   **12.** $\dfrac{3}{10} \times 100 =$ 30

13. What is $\dfrac{2}{3}$ of 48?   32   **14.** What is $\dfrac{1}{5}$ of 90?   18

15. What is $\dfrac{5}{8}$ of 40?   25   **16.** What is $\dfrac{1}{10}$ of 200?   20

17. What is $\dfrac{9}{10}$ of 60?   54   **18.** What is $\dfrac{3}{4}$ of 28?   21

19. One third of 27 is what?   9
20. Two thirds of 36 is what?   24
21. Three fourths of 24 is what?   18
22. Four fifths of 35 is what?   28
23. Two ninths of 36 is what?   8
24. Seven tenths of 30 is what?   21
25. Five twelfths of 24 is what?   10

Set H

Convert these improper fractions to whole numbers or to mixed numbers.

1. $\dfrac{3}{3} = 1$ **2.** $\dfrac{5}{4} = 1\frac{1}{4}$ **3.** $\dfrac{7}{3} = 2\frac{1}{3}$ **4.** $\dfrac{17}{10} = 1\frac{7}{10}$ **5.** $\dfrac{24}{6} = 4$

6. $\dfrac{24}{5} = 4\frac{4}{5}$ **7.** $\dfrac{24}{4} = 6$ **8.** $\dfrac{32}{15} = 2\frac{2}{15}$ **9.** $\dfrac{32}{16} = 2$ **10.** $\dfrac{27}{5} = 5\frac{2}{5}$

11. $\dfrac{36}{7} = 5\frac{1}{7}$ **12.** $\dfrac{25}{6} = 4\frac{1}{6}$ **13.** $\dfrac{35}{5} = 7$ **14.** $\dfrac{12}{5} = 2\frac{2}{5}$ **15.** $\dfrac{31}{10} = 3\frac{1}{10}$

16. $1\dfrac{5}{2} = 3\frac{1}{2}$ **17.** $3\dfrac{6}{3} = 5$ **18.** $7\dfrac{9}{4} = 9\frac{1}{4}$ **19.** $6\dfrac{8}{2} = 10$ **20.** $4\dfrac{5}{3} = 5\frac{2}{3}$

21. $11\dfrac{6}{5} =$ $12\frac{1}{5}$ **22.** $4\dfrac{11}{10} =$ $5\frac{1}{10}$ **23.** $2\dfrac{13}{12} =$ $3\frac{1}{12}$ **24.** $1\dfrac{10}{3} =$ $4\frac{1}{3}$ **25.** $23\dfrac{7}{2} =$ $26\frac{1}{2}$

Set I

Reduce each fraction to lowest terms.

1. $\dfrac{2}{6} = \frac{1}{3}$ 2. $\dfrac{3}{6} = \frac{1}{2}$ 3. $\dfrac{4}{6} = \frac{2}{3}$ 4. $\dfrac{2}{8} = \frac{1}{4}$ 5. $\dfrac{4}{8} = \frac{1}{2}$

6. $\dfrac{6}{8} = \frac{3}{4}$ 7. $\dfrac{3}{9} = \frac{1}{3}$ 8. $\dfrac{2}{10} = \frac{1}{5}$ 9. $\dfrac{4}{10} = \frac{2}{5}$ 10. $\dfrac{5}{10} = \frac{1}{2}$

11. $\dfrac{8}{10} = \frac{4}{5}$ 12. $\dfrac{2}{12} = \frac{1}{6}$ 13. $\dfrac{3}{12} = \frac{1}{4}$ 14. $\dfrac{4}{12} = \frac{1}{3}$ 15. $\dfrac{6}{12} = \frac{1}{2}$

16. $\dfrac{8}{12} = \frac{2}{3}$ 17. $\dfrac{9}{12} = \frac{3}{4}$ 18. $3\dfrac{10}{12} =$ 19. $4\dfrac{6}{15} =$ 20. $1\dfrac{18}{24} =$
                                       $3\frac{5}{6}$ $4\frac{2}{5}$ $1\frac{3}{4}$

21. $2\dfrac{15}{18} =$ 22. $6\dfrac{16}{24} =$ 23. $8\dfrac{12}{24} =$ 24. $9\dfrac{8}{24} =$ 25. $10\dfrac{10}{24} =$
     $2\frac{5}{6}$ $6\frac{2}{3}$ $8\frac{1}{2}$ $9\frac{1}{3}$ $10\frac{5}{12}$

Set J

Simplify.

1. $\dfrac{5}{8} + \dfrac{2}{8} = \frac{7}{8}$ 2. $\dfrac{5}{8} - \dfrac{2}{8} = \frac{3}{8}$ 3. $\dfrac{3}{6} + \dfrac{2}{6} = \frac{5}{6}$

4. $\dfrac{3}{6} - \dfrac{2}{6} = \frac{1}{6}$ 5. $\dfrac{1}{3} + \dfrac{1}{3} = \frac{2}{3}$ 6. $\dfrac{1}{3} - \dfrac{1}{3} = 0$

7. $\dfrac{4}{9} + \dfrac{1}{9} = \frac{5}{9}$ 8. $\dfrac{4}{9} - \dfrac{2}{9} = \frac{2}{9}$ 9. $\dfrac{1}{4} + \dfrac{1}{4} + \dfrac{1}{4} = \frac{3}{4}$

10. $\dfrac{1}{7} + \dfrac{2}{7} + \dfrac{3}{7} = \frac{6}{7}$ 11. $\dfrac{3}{4} + \dfrac{2}{4} = 1\frac{1}{4}$ 12. $\dfrac{3}{4} - \dfrac{1}{4} = \frac{1}{2}$

13. $\dfrac{2}{3} + \dfrac{2}{3} = 1\frac{1}{3}$ 14. $\dfrac{3}{8} - \dfrac{1}{8} = \frac{1}{4}$ 15. $\dfrac{4}{5} + \dfrac{4}{5} = 1\frac{3}{5}$

16. $\dfrac{6}{5} - \dfrac{1}{5} = 1$ 17. $\dfrac{5}{8} + \dfrac{3}{8} = 1$ 18. $\dfrac{5}{8} - \dfrac{1}{8} = \frac{1}{2}$

19. $\dfrac{3}{10} + \dfrac{2}{10} = \frac{1}{2}$ 20. $\dfrac{9}{10} - \dfrac{1}{10} = \frac{4}{5}$ 21. $\dfrac{5}{12} + \dfrac{5}{12} = \frac{5}{6}$

22. $\dfrac{5}{12} - \dfrac{1}{12} = \dfrac{1}{3}$ **23.** $\dfrac{3}{10} + \dfrac{7}{10} = 1$ **24.** $\dfrac{7}{10} - \dfrac{3}{10} = \dfrac{2}{5}$

25. $\dfrac{3}{4} + \dfrac{3}{4} + \dfrac{3}{4} = 2\dfrac{1}{4}$

Set K

Simplify.

1. $3\dfrac{1}{3} + 2\dfrac{1}{3} = 5\dfrac{2}{3}$ **2.** $3\dfrac{1}{3} - 2\dfrac{1}{3} = 1$

3. $5\dfrac{1}{2} + 2 = 7\dfrac{1}{2}$ **4.** $5\dfrac{1}{2} - 2 = 3\dfrac{1}{2}$

5. $7\dfrac{3}{8} + 1\dfrac{2}{8} = 8\dfrac{5}{8}$ **6.** $7\dfrac{3}{8} - 1\dfrac{2}{8} = 6\dfrac{1}{8}$

7. $5\dfrac{1}{4} + \dfrac{2}{4} = 5\dfrac{3}{4}$ **8.** $2\dfrac{1}{4} - 2 = \dfrac{1}{4}$

9. $3\dfrac{7}{10} + 1\dfrac{2}{10} = 4\dfrac{9}{10}$ **10.** $3\dfrac{7}{10} - 2 = 1\dfrac{7}{10}$

11. $1\dfrac{2}{3} + 3\dfrac{1}{3} = 5$ **12.** $4\dfrac{3}{4} - 1\dfrac{1}{4} = 3\dfrac{1}{2}$

13. $5\dfrac{2}{3} + 1\dfrac{2}{3} = 7\dfrac{1}{3}$ **14.** $6\dfrac{5}{8} - 1\dfrac{1}{8} = 5\dfrac{1}{2}$

15. $1\dfrac{3}{4} + 1\dfrac{3}{4} + 1\dfrac{3}{4} = 5\dfrac{1}{4}$ **16.** $5\dfrac{7}{10} - 1\dfrac{1}{10} = 4\dfrac{3}{5}$

17. $1\dfrac{3}{5} + \dfrac{2}{5} = 2$ **18.** $6\dfrac{1}{2} - 4\dfrac{1}{2} = 2$

19. $5\dfrac{8}{9} + 2\dfrac{5}{9} = 8\dfrac{4}{9}$ **20.** $5\dfrac{8}{9} - 2\dfrac{5}{9} = 3\dfrac{1}{3}$

21. $\dfrac{4}{5} + \dfrac{4}{5} + \dfrac{4}{5} = 2\dfrac{2}{5}$ **22.** $1\dfrac{8}{8} - 1\dfrac{1}{8} = \dfrac{7}{8}$

23. $3\dfrac{1}{2} + 3\dfrac{1}{2} + 3\dfrac{1}{2} = 10\dfrac{1}{2}$ **24.** $3\dfrac{3}{2} - 1\dfrac{1}{2} = 3$

25. $4\dfrac{5}{4} - 1\dfrac{3}{4} = 3\dfrac{1}{2}$

Set L

Find the Least Common Multiple (LCM) of each set of numbers.
1. 3, 4   12   **2.** 3, 5   15   **3.** 3, 6   6   **4.** 4, 6   12
5. 6, 8   24   **6.** 4, 8   8   **7.** 3, 8   24   **8.** 2, 8   8
9. 3, 9   9   **10.** 6, 9   18   **11.** 6, 10   30   **12.** 4, 10   20
13. 8, 12   24   **14.** 9, 12   36   **15.** 10, 12   60   **16.** 2, 5, 10   10
17. 2, 3, 4   12   **18.** 2, 3, 6   6   **19.** 2, 4, 8   8   **20.** 2, 4, 6   12
21. 3, 4, 6   12   **22.** 6, 8, 12   24   **23.** 5, 10, 15   30   **24.** 10, 20, 30
25. 20, 30, 40   120   60

Set M

Write the standard numeral for each number.
1. $(6 \times 100) + (7 \times 10)$ 670   **2.** $(5 \times 1000) + (4 \times 100)$ 5400
3. $(7 \times 100) + (3 \times 1)$ 703   **4.** $(8 \times 10) + (1 \times 1)$ 81
5. $(9 \times 1000) + (5 \times 10)$ 9050   **6.** $(7 \times 100) + (3 \times 10)$ 730
7. $(5 \times 100) + (3 \times 1)$ 503   **8.** $(3 \times 1) + (7 \times \frac{1}{10})$ 3.7
9. $(6 \times \frac{1}{10}) + (2 \times \frac{1}{100})$ 0.62   **10.** $(5 \times 10) + (4 \times \frac{1}{10})$ 50.4
11. $(3 \times 1) + (2 \times \frac{1}{100})$ 3.02   **12.** $(8 \times 10) + (4 \times \frac{1}{100})$ 80.04
13. $(9 \times 10) + (6 \times \frac{1}{10})$ 90.6   **14.** $(4 \times \frac{1}{100})$ 0.04
15. $(8 \times 10) + (8 \times \frac{1}{100})$ 80.04

Write each number in expanded notation.
16. 560   **17.** 300   **18.** 5600   **19.** 706
$(5 \times 100) + (6 \times 10)$ 3×100 $(5 \times 1000) + (6 \times 100)$ $(7 \times 100) + (6 \times 1)$
20. 4.8   **21.** .25   **22.** .05   **23.** 20.6
$(4 \times 1) + (8 \times \frac{1}{10})$ $(2 \times \frac{1}{10}) + (5 \times \frac{1}{100})$ $5 \times \frac{1}{100}$ $(2 \times 10) + (6 \times \frac{1}{10})$
24. 5280 $(5 \times 1000) + (2 \times 100) + (8 \times 10)$ **25.** 6.24 $(6 \times 1) + (2 \times \frac{1}{10}) + (4 \times \frac{1}{100})$

Set N

Write the decimal numeral named.
1. Five tenths   0.5   **4.** One thousandth   0.001
2. Three hundredths   0.03   **5.** Twenty-five thousandths   0.025
3. Eleven hundredths   0.11   **6.** One and two tenths   1.2

7. Ten and four tenths   10.4
8. Two and one hundredth   2.01
9. Five and twelve hundredths   5.12
10. One hundred twenty thousandths   0.120
11. Two hundred five thousandths   0.205
12. Six and fifteen hundredths   6.15
13. Ten and one hundred thousandths   10.100
14. Twelve and six hundredths   12.06
15. Ten and twenty-two thousandths   10.022

Write the decimal form of each fraction or mixed number.

16. $\dfrac{5}{100}$ 0.05
17. $\dfrac{12}{1000}$ 0.012
18. $1\dfrac{3}{10}$ 1.3
19. $10\dfrac{1}{10}$ 10.1
20. $5\dfrac{23}{100}$ 5.23

21. $\dfrac{124}{1000}$ 0.124
22. $1\dfrac{1}{1000}$ 1.001
23. $1\dfrac{45}{100}$ 1.45
24. $8\dfrac{3}{100}$ 8.03
25. $9\dfrac{52}{1000}$ 9.052

Set O

1. .62 + .4 = 1.02
2. .62 − .4 = 0.22
3. 1.5 + .15 = 1.65
4. 1.2 − .15 = 1.05
5. .5 + .41 = 0.91
6. .5 − .41 = 0.09
7. .23 + .6 + 1.4 = 2.23
8. 5.3 − 4.29 = 1.01
9. 3.6 + 2 + .75 = 6.35
10. 1 − .3 = 0.7
11. 4.75 + 3 + 12.5 = 20.25
12. 15.4 − 15.40 = 0
13. .3 + .4 + .5 = 1.2
14. 5 − 1.25 = 3.75
15. .36 + .4 + .575 = 1.335
16. .3 − .036 = 0.264
17. 1 + .2 + 3.456 = 4.656
18. 10 − .7 = 9.3
19. .6 + .7 + .8 = 2.1
20. 1 − .21 = 0.79
21. .8 + 8 + 8.88 = 17.68
22. 2.34 − .5 = 1.84
23. 1.28 + .4 + 7.6 = 9.28
24. 18.3 − 7.924 = 10.376
25. 6.78 + 6 + .78 = 13.56

Set P

1. .3 × 4 = 1.2
2. .4 × .6 = 0.24
3. .3 × .2 = 0.06
4. .4 × .3 × .2 = 0.024
5. 7 × .21 = 1.47
6. .6 × 1.24 = 0.744
7. .36 × .4 = 0.144
8. .012 × 10 = 0.12
9. 1.2 × 8 = 9.6

10. $6.2 \times .07 =$ 0.434   **11.** $1.2 \times .12 =$ 0.144   **12.** $1.25 \times 10 =$ 12.5

13. $3.6 \times 1.2 =$ 4.32   **14.** $4.5 \times 9 =$ 40.5   **15.** $.015 \times .03 =$ 0.00045

16. $6.75 \times .1 =$ 0.675   **17.** $.01 \times 3.75 =$ 0.0375   **18.** $1.5 \times 1.5 =$ 2.25

19. $.25 \times .25 =$ 0.0625   **20.** $6.3 \times .24 =$ 1.512   **21.** $4.2 \times 100 =$ 420

22. $.15 \times .013 =$ 0.00195   **23.** $42 \times .16 =$ 6.72   **24.** $7.2 \times .24 =$ 1.728

25. $.1 \times .01 \times 1 =$ 0.001

Set Q

1. $4.8 \div 6 =$ 0.8   **2.** $.48 \div 4 =$ 0.12   **3.** $.48 \div 8 =$ 0.06

4. $.125 \div 5 =$ 0.025   **5.** $1.44 \div 6 =$ 0.24   **6.** $.018 \div 3 =$ 0.006

7. $.24 \div 12 =$ 0.02   **8.** $5.6 \div 8 =$ 0.7   **9.** $17.1 \div 9 =$ 1.9

10. $3.65 \div 5 =$ 0.73   **11.** $42.80 \div 10 =$ 4.28   **12.** $3.10 \div 10 =$ 0.31

13. $.190 \div 5 =$ 0.038   **14.** $.234 \div 9 =$ 0.026   **15.** $5.00 \div 4 =$ 1.25

16. $.7 \div 5 =$ 0.14   **17.** $.4 \div 4 =$ 0.1   **18.** $.5 \div 4 =$ 0.125

19. $3.6 \div 10 =$ 0.36   **20.** $.24 \div 10 =$ 0.024   **21.** $.12 \div 8 =$ 0.015

22. $.9 \div 4 =$ 0.225   **23.** $1.1 \div 8 =$ 0.1375   **24.** $.51 \div 10 =$ 0.051

25. $4.32 \div 12 =$ 0.36

Set R

1. $5.2 \div .4 =$ 13   **2.** $.144 \div .8 =$ 0.18   **3.** $3.21 \div .3 =$ 10.7

4. $1.00 \div .4 =$ 2.5   **5.** $.525 \div .05 =$ 10.5   **6.** $8.1 \div .09 =$ 90

7. $1.2 \div .003 =$ 400   **8.** $.54 \div .006 =$ 90   **9.** $1.2 \div .12 =$ 10

10. $.12 \div 1.2 =$ 0.1   **11.** $.5 \div .04 =$ 12.5   **12.** $3.6 \div .5 =$ 7.2

13. $.12 \div 10 =$ 0.012   **14.** $6.4 \div 100 =$ 0.064   **15.** $3.5 \div .08 =$ 43.75

16. $3 \div .5 =$ 16   **17.** $4 \div .4 =$ 10   **18.** $12 \div .06 =$ 200

19. $18 \div .20 =$ 90   **20.** $16 \div .008 =$ 2000   **21.** $5 \div .25 =$ 20

22. $4.44 \div .06 =$ 74   **23.** $16 \div .25 =$ 64   **24.** $.3 \div .4 =$ 0.75

25. $1 \div .08 =$ 12.5

Set S

Round each to the nearest tenth.

1. .48   0.5   **2.** .133   0.1   **3.** .375   0.4   **4.** 4.28   4.3

5. 62.84   62.8   **6.** .0984   0.1   **7.** 6.25   6.3   **8.** 1.97   2.0

Round each to the nearest hundredth (or cent).
9. .8181   0.82   **10.** .6666   0.67   **11.** 1.333   1.33   **12.** 4.321   4.32
13. .2345   0.23   **14.** 7.675   7.68   **15.** $0.166   **16.** $3.422
                                              $0.17 $3.42
17. $15.555   **18.** $1.8975
    $15.56 $1.90

Round each to the nearest thousandth.
19. .1234   **20.** .4567   **21.** .3542   **22.** .9009
    0.123   0.457   0.354   0.901
23. .0833   **24.** .9166   **25.** .142857
    0.083   0.917   0.143

Set T

Solve mentally.
1. $4.2 \times 10 = 42$ **2.** $.35 \times 10 = 3.5$
3. $.178 \times 10 = 1.78$ **4.** $3.65 \times 10 = 36.5$
5. $4.21 \times 100 = 421$ **6.** $.375 \times 100 = 37.5$
7. $6.5 \times 100 = 650$ **8.** $4.323 \times 100 = 432.3$
9. $7.275 \times 1000 = 7275$ **10.** $6.4 \times 1000 = 6400$
11. $.86 \times 1000 = 860$ **12.** $.01625 \times 1000 = 16.25$
13. $4.2 \div 10 = 0.42$ **14.** $.42 \div 10 = 0.042$
15. $42.1 \div 10 = 4.21$ **16.** $6 \div 10 = 0.6$
17. $87.5 \div 100 = 0.875$ **18.** $6.5 \div 100 = 0.065$
19. $.4 \div 100 = 0.004$ **20.** $372.8 \div 100 = 3.728$
21. $123.4 \div 1000 = 0.1234$ **22.** $42.5 \div 1000 = 0.0425$
23. $7.6 \div 1000 = 0.0076$ **24.** $4 \div 1000 = 0.004$
25. $.3 \div 1000 = 0.0003$

Set U

1. $1 - \frac{1}{5} = \frac{4}{5}$ **2.** $1 - \frac{3}{8} = \frac{5}{8}$ **3.** $2 - \frac{1}{2} = 1\frac{1}{2}$

4. $2 - \frac{1}{3} = 1\frac{2}{3}$ **5.** $2 - 1\frac{1}{4} = \frac{3}{4}$ **6.** $3 - 1\frac{3}{8} = 1\frac{5}{8}$

7. $3 - 2\frac{5}{8} = \frac{3}{8}$ **8.** $8 - 3\frac{3}{4} = 4\frac{1}{4}$ **9.** $8 - 5\frac{1}{8} = 2\frac{7}{8}$

10. $10 - 4\dfrac{2}{5} = 5\dfrac{3}{5}$ **11.** $4\dfrac{1}{3} - 1\dfrac{2}{3} = 2\dfrac{2}{3}$ **12.** $4\dfrac{1}{5} - 1\dfrac{3}{5} = 2\dfrac{3}{5}$

13. $6\dfrac{1}{10} - 3\dfrac{4}{10} = 2\dfrac{7}{10}$ **14.** $5\dfrac{3}{8} - 3\dfrac{6}{8} = 1\dfrac{5}{8}$ **15.** $2\dfrac{1}{4} - 1\dfrac{2}{4} = \dfrac{3}{4}$

16. $5\dfrac{2}{5} - 3\dfrac{4}{5} = 1\dfrac{3}{5}$ **17.** $7\dfrac{3}{8} - 4\dfrac{4}{8} = 2\dfrac{7}{8}$ **18.** $9\dfrac{1}{10} - 3\dfrac{5}{10} = 5\dfrac{3}{5}$

19. $4\dfrac{2}{4} - 1\dfrac{3}{4} = 2\dfrac{3}{4}$ **20.** $8\dfrac{3}{5} - 1\dfrac{4}{5} = 6\dfrac{4}{5}$ **21.** $4\dfrac{1}{4} - 1\dfrac{3}{4} = 2\dfrac{1}{2}$

22. $4\dfrac{3}{8} - 1\dfrac{5}{8} = 2\dfrac{3}{4}$ **23.** $3\dfrac{1}{6} - 1\dfrac{5}{6} = 1\dfrac{1}{3}$ **24.** $6\dfrac{1}{10} - 4\dfrac{3}{10} = 1\dfrac{4}{5}$

25. $2\dfrac{1}{12} - 1\dfrac{3}{12} = \dfrac{5}{6}$

Set V

1. $\dfrac{1}{2} + \dfrac{1}{8} = \dfrac{5}{8}$ **2.** $\dfrac{1}{2} - \dfrac{1}{8} = \dfrac{3}{8}$ **3.** $\dfrac{3}{4} + \dfrac{1}{8} = \dfrac{7}{8}$

4. $\dfrac{3}{4} - \dfrac{1}{8} = \dfrac{5}{8}$ **5.** $\dfrac{2}{3} + \dfrac{1}{6} = \dfrac{5}{6}$ **6.** $\dfrac{2}{3} - \dfrac{1}{6} = \dfrac{1}{2}$

7. $\dfrac{1}{3} + \dfrac{1}{4} = \dfrac{7}{12}$ **8.** $\dfrac{1}{3} - \dfrac{1}{4} = \dfrac{1}{12}$ **9.** $\dfrac{3}{4} + \dfrac{2}{3} = 1\dfrac{5}{12}$

10. $\dfrac{3}{4} - \dfrac{2}{3} = \dfrac{1}{12}$ **11.** $\dfrac{1}{2} + \dfrac{1}{10} = \dfrac{3}{5}$ **12.** $\dfrac{1}{2} - \dfrac{1}{10} = \dfrac{2}{5}$

13. $\dfrac{3}{4} + \dfrac{3}{8} = 1\dfrac{1}{8}$ **14.** $\dfrac{3}{4} - \dfrac{3}{8} = \dfrac{3}{8}$ **15.** $\dfrac{2}{3} + \dfrac{1}{2} = 1\dfrac{1}{6}$

16. $\dfrac{2}{3} - \dfrac{1}{2} = \dfrac{1}{6}$ **17.** $\dfrac{7}{10} + \dfrac{1}{2} = 1\dfrac{1}{5}$ **18.** $\dfrac{7}{10} - \dfrac{1}{2} = \dfrac{1}{5}$

19. $\dfrac{1}{4} + \dfrac{1}{5} = \dfrac{9}{20}$ **20.** $\dfrac{1}{4} - \dfrac{1}{5} = \dfrac{1}{20}$ **21.** $\dfrac{3}{5} + \dfrac{1}{2} = 1\dfrac{1}{10}$

22. $\dfrac{3}{5} - \dfrac{1}{2} = \dfrac{1}{10}$ **23.** $\dfrac{1}{2} + \dfrac{1}{4} + \dfrac{1}{8} = \dfrac{7}{8}$ **24.** $\dfrac{1}{3} + \dfrac{1}{4} + \dfrac{1}{6} = \dfrac{3}{4}$

25. $\dfrac{1}{2} + \dfrac{2}{3} + \dfrac{3}{4} = 1\dfrac{11}{12}$

Set W

1. $3\dfrac{1}{8} + 2\dfrac{1}{4} = 5\dfrac{3}{8}$ **2.** $3\dfrac{1}{4} - 2\dfrac{1}{8} = 1\dfrac{1}{8}$ **3.** $1\dfrac{1}{6} + 1\dfrac{1}{3} = 2\dfrac{1}{2}$

4. $2\dfrac{1}{3} - 1\dfrac{1}{6} = 1\dfrac{1}{6}$ **5.** $3\dfrac{3}{4} + 4\dfrac{1}{8} = 7\dfrac{7}{8}$ **6.** $4\dfrac{3}{4} - 3\dfrac{1}{8} = 1\dfrac{5}{8}$

7. $5\dfrac{3}{5} + 1\dfrac{3}{10} = 6\dfrac{9}{10}$ **8.** $5\dfrac{3}{5} - 1\dfrac{3}{10} = 4\dfrac{3}{10}$ **9.** $4\dfrac{1}{2} + 2\dfrac{1}{12} = 6\dfrac{7}{12}$

10. $4\dfrac{1}{2} - 2\dfrac{1}{12} = 2\dfrac{5}{12}$ **11.** $6\dfrac{1}{2} + 2\dfrac{1}{3} = 8\dfrac{5}{6}$ **12.** $4\dfrac{1}{2} - 2\dfrac{1}{3} = 2\dfrac{1}{6}$

13. $5\dfrac{1}{2} + 1\dfrac{2}{3} = 7\dfrac{1}{6}$ **14.** $5\dfrac{1}{2} - 1\dfrac{2}{3} = 3\dfrac{5}{6}$ **15.** $1\dfrac{1}{2} + \dfrac{3}{4} = 2\dfrac{1}{4}$

16. $1\dfrac{1}{2} - \dfrac{3}{4} = \dfrac{3}{4}$ **17.** $6\dfrac{3}{5} + 1\dfrac{1}{2} = 8\dfrac{1}{10}$ **18.** $6\dfrac{3}{5} - 1\dfrac{1}{2} = 5\dfrac{1}{10}$

19. $2\dfrac{7}{10} + 1\dfrac{1}{5} = 3\dfrac{9}{10}$ **20.** $2\dfrac{7}{10} - 1\dfrac{1}{5} = 1\dfrac{1}{2}$ **21.** $8\dfrac{2}{3} + 1\dfrac{3}{4} = 10\dfrac{5}{12}$

22. $8\dfrac{2}{3} - 1\dfrac{3}{4} = 6\dfrac{11}{12}$ **23.** $3\dfrac{1}{2} + 1\dfrac{1}{3} + 2\dfrac{1}{6} = 7$ **24.** $5\dfrac{1}{4} + 3\dfrac{1}{2} + 1\dfrac{3}{8} = 10\dfrac{1}{8}$

25. $7\dfrac{1}{2} + 2\dfrac{2}{3} + 3\dfrac{1}{4} = 13\dfrac{5}{12}$

Set X

Write the prime factorization of each number.

1. 6 $\quad 2 \times 3$ **2.** 7 $\quad 7$ **3.** 8 $\quad 2 \times 2 \times 2$ **4.** 9 $\quad 3 \times 3$ **5.** 10 $\quad 2 \times 5$

6. 11 $\quad 11$ **7.** 12 $\quad 2 \times 2 \times 3$ **8.** 13 $\quad 13$ **9.** 14 $\quad 2 \times 7$ **10.** 15 $\quad 3 \times 5$

11. 16 $\quad 2 \times 2 \times 2 \times 2$ **12.** 18 $\quad 2 \times 3 \times 3$ **13.** 20 $\quad 2 \times 2 \times 5$ **14.** 21 $\quad 3 \times 7$ **15.** 24 $\quad 2 \times 2 \times 2 \times 3$

16. 30 $\quad 2 \times 3 \times 5$ **17.** 36 $\quad 2 \times 2 \times 3 \times 3$ **18.** 39 $\quad 3 \times 13$ **19.** 40 $\quad 2 \times 2 \times 2 \times 5$ **20.** 41 $\quad 41$

21. 42 $\quad 2 \times 3 \times 7$ **22.** 48 $\quad 2 \times 2 \times 2 \times 2 \times 3$ **23.** 60 $\quad 2 \times 2 \times 3 \times 5$ **24.** 100 $\quad 2 \times 2 \times 5 \times 5$ **25.** 144 $\quad 2 \times 2 \times 2 \times 2 \times 3 \times 3$

Set Y

1. $3 \times 1\dfrac{1}{4} = 3\dfrac{3}{4}$ **2.** $1\dfrac{1}{2} \times 3 = 4\dfrac{1}{2}$ **3.** $1\dfrac{1}{2} \times 1\dfrac{1}{4} = 1\dfrac{7}{8}$

4. $1\frac{2}{3} \times 2\frac{1}{2} = 4\frac{1}{6}$ **5.** $3\frac{1}{2} \times 5 = 17\frac{1}{2}$ **6.** $1\frac{3}{4} \times 1\frac{1}{2} = 2\frac{5}{8}$

7. $3\frac{1}{3} \times 1\frac{2}{3} = 5\frac{5}{9}$ **8.** $7\frac{1}{2} \times 2 = 15$ **9.** $\frac{4}{5} \times 1\frac{1}{5} = \frac{24}{25}$

10. $\frac{5}{6} \times 1\frac{1}{5} = 1$ **11.** $1\frac{1}{2} \times 1\frac{1}{3} = 2$ **12.** $1\frac{1}{2} \times 1\frac{2}{3} = 2\frac{1}{2}$

13. $1\frac{1}{4} \times 2\frac{2}{5} = 3$ **14.** $3\frac{2}{3} \times 3 = 11$ **15.** $4 \times 3\frac{1}{2} = 14$

16. $\frac{5}{6} \times 3\frac{3}{5} = 3$ **17.** $3\frac{1}{3} \times 2\frac{1}{10} = 7$ **18.** $5\frac{1}{3} \times 1\frac{1}{8} = 6$

19. $2\frac{1}{2} \times 1\frac{1}{3} = 3\frac{1}{3}$ **20.** $\frac{7}{8} \times 2\frac{2}{3} = 2\frac{1}{3}$ **21.** $1\frac{2}{5} \times 2\frac{1}{2} = 3\frac{1}{2}$

22. $4\frac{1}{2} \times 2\frac{2}{3} = 12$ **23.** $5 \times 1\frac{3}{5} \times \frac{3}{8} = 3$ **24.** $\frac{5}{8} \times 3\frac{1}{5} \times \frac{1}{2} = 1$

25. $1\frac{1}{2} \times 1\frac{2}{3} \times 1\frac{1}{5} = 3$

Set Z

1. $1\frac{1}{2} \div 3 = \frac{1}{2}$ **2.** $3 \div 1\frac{1}{2} = 2$ **3.** $1\frac{2}{3} \div 2 = \frac{5}{6}$

4. $2 \div 1\frac{2}{3} = 1\frac{1}{5}$ **5.** $\frac{3}{4} \div 1\frac{1}{2} = \frac{1}{2}$ **6.** $1\frac{1}{2} \div \frac{3}{4} = 2$

7. $1\frac{2}{3} \div 1\frac{1}{2} = 1\frac{1}{9}$ **8.** $1\frac{1}{2} \div 1\frac{2}{3} = \frac{9}{10}$ **9.** $\frac{3}{8} \div 2 = \frac{3}{16}$

10. $2 \div \frac{3}{8} = 5\frac{1}{3}$ **11.** $1\frac{3}{5} \div 2\frac{1}{3} = \frac{24}{35}$ **12.** $2\frac{1}{3} \div 1\frac{3}{5} = 1\frac{11}{24}$

13. $4\frac{1}{2} \div 2\frac{1}{4} = 2$ **14.** $2\frac{1}{4} \div 4\frac{1}{2} = \frac{1}{2}$ **15.** $5 \div 1\frac{1}{4} = 4$

16. $1\frac{1}{4} \div 5 = \frac{1}{4}$ **17.** $2\frac{2}{3} \div 2 = 1\frac{1}{3}$ **18.** $2 \div 2\frac{2}{3} = \frac{3}{4}$

19. $2\frac{1}{2} \div 1\frac{3}{4} = 1\frac{3}{7}$ **20.** $1\frac{3}{4} \div 2\frac{1}{2} = \frac{7}{10}$ **21.** $\frac{3}{4} \div 2\frac{1}{4} = \frac{1}{3}$

22. $2\dfrac{1}{4} \div \dfrac{3}{4} = 3$ **23.** $3\dfrac{1}{3} \div 2\dfrac{1}{2} = 1\dfrac{1}{3}$ **24.** $2\dfrac{1}{2} \div 3\dfrac{1}{3} = \dfrac{3}{4}$

25. $6\dfrac{2}{3} \div 6\dfrac{2}{3} = 1$

Set AA

Compare.

1. .3 ◯ .12   >   **2.** .01 ◯ .011   <   **3.** .204 ◯ .24   <

4. 3.5 ◯ .36   >   **5.** .51 ◯ .509   >   **6.** 4.6 ◯ 4.601   <

7. .35 ◯ .315   >   **8.** .7 ◯ .700   =   **9.** .49 ◯ .051   >

10. .1 ◯ .099   >

Arrange in order of size from least to greatest.

11. .3, .31, .301
0.3, 0.301, 0.31

12. .6, .06, 6.0
0.06, 0.6, 6.0

13. 7.2, .72, 7.02
0.72, 7.02, 7.2

14. .41, .041, .401
0.041, 0.401, 0.41

15. .316, .32, .31
0.31, 0.316, 0.32

16. .1, .09, .099
0.09, 0.099, 0.1

17. 6.01, 0.61, 6.10
0.61, 6.01, 6.10

18. 4.03, .403, .043
0.043, 0.403, 4.03

19. 3.75, 37.5, 37.05
3.75, 37.05, 37.5

20. .02, .002, .2   0.002, 0.02, 0.2

Find the number in each set that is closest to 1.

21. .5, 5, .05   0.5   **22.** .1, 1.1, 1.11   1.1   **23.** .9, 2, 1.9   0.9

24. 0, 2, .4   0.4   **25.** 0, 2.1, 3   0

Set BB

Write each fraction as a decimal.

1. $\dfrac{1}{2} = 0.5$ **2.** $\dfrac{1}{4} = 0.25$ **3.** $\dfrac{1}{8} = 0.125$ **4.** $\dfrac{1}{10} = 0.1$ **5.** $\dfrac{3}{4} = 0.75$

6. $\dfrac{3}{8} = 0.375$ **7.** $\dfrac{3}{10} = 0.3$ **8.** $\dfrac{3}{5} = 0.6$ **9.** $\dfrac{5}{8} = 0.625$ **10.** $\dfrac{7}{10} = 0.7$

11. $\dfrac{4}{5} = 0.8$ **12.** $\dfrac{3}{20} = 0.15$ **13.** $\dfrac{7}{50} = 0.14$

Write each decimal as a reduced fraction.

14. 0.1 $\frac{1}{10}$ **15.** 0.2 $\frac{1}{5}$ **16.** 0.4 $\frac{2}{5}$ **17.** 0.5 $\frac{1}{2}$

18. 0.8 $\frac{4}{5}$ **19.** 0.9 $\frac{9}{10}$ **20.** 0.12 $\frac{3}{25}$ **21.** 0.15 $\frac{3}{20}$

22. 0.25 $\frac{1}{4}$ **23.** 0.75 $\frac{3}{4}$ **24.** 0.025 $\frac{1}{40}$ **25.** 0.005 $\frac{1}{200}$

Set CC

Two fifths of the 25 ballplayers played.

1. Into how many parts was the team divided?   5

2. How many are in each part?   5

3. How many parts played?   2

4. How many players played?   10

5. How many parts did not play?   3

6. How many players did not play?   15

Five sixths of the 300 members had paid their dues.

7. Into how many parts was the group divided?   6

8. How many members are in each part?   50

9. How many parts paid dues?   5

10. How many members paid their dues?   250

11. How many parts did not pay dues?   1

12. How many members did not pay their dues?   50

Three tenths of the $6000 in prize money went to pay taxes.

13. Into how many parts was the prize money divided?   10

14. How much money was in each part? $600

15. How many parts went to paying taxes?   3

16. How much money went to paying taxes? $1800

17. How many parts did not go to paying taxes?   7

18. How much money did not go to paying taxes? $4200

Out of 800 students, $\frac{27}{100}$ rode their bikes to school.

19. Into how many parts were the students divided?   100

20. How many students were in each part?   8

21. How many parts rode bikes?   27

22. How many students rode their bikes?   216

23. How many parts did not ride bikes?   73
24. How many students did not ride their bikes?   584

Set DD

Thirty percent of the 200 passengers sat alone.
1. Into how many parts were the passengers divided?   100
2. How many were in each part?   2
3. How many parts sat alone?   30
4. How many passengers sat alone?   60
5. How many parts did not sit alone?   70
6. How many passengers did not sit alone?   140

Seventy-five percent of the one thousand trees cost more than $20.
7. Into how many parts are the trees divided?   100
8. How many trees are in each part?   10
9. How many parts cost more than $20?   75
10. How many trees cost more than $20?   750
11. How many parts do not cost more than $20?   25
12. How many trees do not cost more than $20?   250

Forty percent of the $5.00 price is profit.
13. Into how many parts is the $5.00 divided?   100
14. What is the value of each part?   5¢
15. How many parts are profit?   40
16. What amount of money is the profit? $2.00
17. How many parts are not profit?   60
18. What amount of money is not profit? $3.00

Eighty percent of the ten thousand seeds are expected to sprout.
19. Into how many parts are the seeds divided?   100
20. How many seeds are in each part?   100
21. How many parts are expected to sprout?   80
22. How many parts are not expected to sprout?   20
23. How many seeds are expected to sprout?   8000
24. How many seeds are not expected to sprout?   2000

Set EE

	PERCENT		FRACTION		DECIMAL
1.	10%	**a.**	$\frac{1}{10}$	**b.**	0.1
2.	20%	**a.**	$\frac{1}{5}$	**b.**	0.2
3.	30%	**a.**	$\frac{3}{10}$	**b.**	0.3
4.	40%	**a.**	$\frac{2}{5}$	**b.**	0.4
5.	50%	**a.**	$\frac{1}{2}$	**b.**	0.5
6.	15%	**a.**	$\frac{3}{20}$	**b.**	0.15
7.	25%	**a.**	$\frac{1}{4}$	**b.**	0.25
8.	45%	**a.**	$\frac{9}{20}$	**b.**	0.45
9.	75%	**a.**	$\frac{3}{4}$	**b.**	0.75
10.	1%	**a.**	$\frac{1}{100}$	**b.**	0.01
11.	2%	**a.**	$\frac{1}{50}$	**b.**	0.02
12.	4%	**a.**	$\frac{1}{25}$	**b.**	0.04
13.	5%	**a.**	$\frac{1}{20}$	**b.**	0.05
14.	6%	**a.**	$\frac{3}{50}$	**b.**	0.06
15.	12%	**a.**	$\frac{3}{25}$	**b.**	0.12
16.	24%	**a.**	$\frac{6}{25}$	**b.**	0.24
17.	90%	**a.**	$\frac{9}{10}$	**b.**	0.9
18	95%	**a.**	$\frac{19}{20}$	**b.**	0.95
19	36%	**a.**	$\frac{9}{25}$	**b.**	0.36
20.	150%	**a.**	$1\frac{1}{2}$	**b.**	1.5
21.	250%	**a.**	$2\frac{1}{2}$	**b.**	2.5
22.	110%	**a.**	$1\frac{1}{10}$	**b.**	1.1
23.	125%	**a.**	$1\frac{1}{4}$	**b.**	1.25
24.	120%	**a.**	$1\frac{1}{5}$	**b.**	1.2
25.	105%	**a.**	$1\frac{1}{20}$	**b.**	1.05

Set FF

Write each answer as a fraction.

1. $0.5 + \dfrac{1}{3} = \dfrac{5}{6}$ **2.** $0.8 - \dfrac{2}{5} = \dfrac{2}{5}$ **3.** $\dfrac{3}{4} - 0.1 = \dfrac{13}{20}$

4. $\dfrac{1}{3} \times 0.6 = \dfrac{1}{5}$ **5.** $0.2 \times \dfrac{1}{2} = \dfrac{1}{10}$ **6.** $0.75 \div \dfrac{3}{4} = 1$

7. $\dfrac{3}{4} \div 0.25 = 3$ **8.** $\dfrac{7}{10} + 0.3 = 1$ **9.** $\dfrac{1}{10} + 0.3 = \dfrac{2}{5}$

10. $0.4 - \dfrac{3}{10} = \dfrac{1}{10}$ **11.** $\dfrac{2}{3} \times 0.75 = \dfrac{1}{2}$ **12.** $\dfrac{1}{5} \div 0.6 = \dfrac{1}{3}$

Write each answer as a decimal.

13. $3.6 + \dfrac{1}{2} = 4.1$ **14.** $\dfrac{1}{4} - 0.2 = 0.05$ **15.** $1.2 \times \dfrac{1}{5} = 0.24$

16. $\dfrac{3}{5} \div 0.3 = 2$ **17.** $4.4 + \dfrac{3}{5} = 5$ **18.** $\dfrac{3}{4} - 0.15 = 0.6$

19. $1.2 \times \dfrac{3}{4} = 0.9$ **20.** $\dfrac{3}{10} \div 0.3 = 1$ **21.** $1 + \dfrac{1}{2} = 1.5$

22. $\dfrac{3}{8} - 0.3 = 0.075$ **23.** $0.8 \times \dfrac{1}{8} = 0.1$ **24.** $\dfrac{1}{2} \div 0.2 = 2.5$

25. $1.2 + \dfrac{3}{4} + 0.05 = 2$

Set GG

Write each fraction as a percent.

1. $\dfrac{1}{10}$ 10% **2.** $\dfrac{9}{10}$ 90% **3.** $\dfrac{1}{5}$ 20% **4.** $\dfrac{3}{4}$ 75% **5.** $\dfrac{3}{20}$ 15%

6. $\dfrac{3}{25}$ 12% **7.** $\dfrac{3}{50}$ 6% **8.** $\dfrac{3}{100}$ 3% **9.** $\dfrac{11}{100}$ 11% **10.** $\dfrac{11}{50}$ 22%

11. $\dfrac{11}{25}$ 44% **12.** $\dfrac{11}{20}$ 55% **13.** $\dfrac{1}{3}$ $33\tfrac{1}{3}$% **14.** $\dfrac{2}{3}$ $66\tfrac{2}{3}$% **15.** $\dfrac{1}{8}$ $12\tfrac{1}{2}$%

16. $\frac{3}{8}$ $37\frac{1}{2}\%$ **17.** $\frac{1}{9}$ $11\frac{1}{9}\%$ **18.** $\frac{5}{4}$ 125% **19.** $\frac{1}{6}$ $16\frac{2}{3}\%$ **20.** $\frac{5}{8}$ $62\frac{1}{2}\%$

21. $\frac{1}{7}$ $14\frac{2}{7}\%$ **22.** $\frac{5}{2}$ 250% **23.** $\frac{5}{6}$ $83\frac{1}{3}\%$ **24.** $\frac{7}{9}$ $77\frac{7}{9}\%$ **25.** $\frac{5}{12}$ $41\frac{2}{3}\%$

Set HH

Write each decimal as a percent.

1. 0.6   60% **2.** 3.4 340% **3.** 0.01   1% **4.** 1.2 120% **5.** 0.5   50%
6. 1.0 100% **7.** 0.37 37% **8.** 4.5 450% **9.** 2.0   **10.** 0.1 10%
11. 1.05   **12.** 0.6 60% **13.** 3.0 300%   200%
105%

Write each percent as a decimal.

14. 25% 0.25 **15.** 20%   0.2 **16.** 120%   1.2 **17.** 125% 1.25
18. 3% 0.03 **19.** 70% 0.7 **20.** 1%   0.01 **21.** 200% 2
22. 6% 0.06 **23.** 24% 0.24 **24.** 375% 3.75 **25.** 9% 0.09

Set II

	FRACTION	DECIMAL	PERCENT
1.	$\frac{1}{100}$	**a.** 0.01	**b.** 1%
2.	**a.** $\frac{4}{5}$	0.8	**b.** 80%
3.	**a.** $\frac{1}{4}$	**b.** 0.25	25%
4.	$\frac{3}{4}$	**a.** 0.75	**b.** 75%

	FRACTION	DECIMAL	PERCENT
5.	a. $\frac{7}{10}$	0.7	b. 70%
6.	a. $\frac{9}{10}$	b. 0.9	90%
7.	$\frac{1}{20}$	a. 0.05	b. 5%
8.	a. $\frac{1}{2}$	0.5	b. 50%
9.	a. $\frac{1}{25}$	b. 0.04	4%
10.	$\frac{1}{50}$	a. 0.02	b. 2%
11.	a. $\frac{9}{20}$	0.45	b. 45%
12.	a. $\frac{23}{100}$	b. 0.23	23%
13.	$\frac{3}{4}$	a. 0.75	b. 75%
14.	a. $\frac{3}{20}$	0.15	b. 15%
15.	a. $\frac{1}{10}$	b. 0.1	10%
16.	$\frac{1}{8}$	a. 0.125	b. $12\frac{1}{2}$%
17.	a. $\frac{1}{5}$	0.2	b. 20%

	FRACTION	DECIMAL	PERCENT
18.	**a.** $\frac{7}{20}$	**b.** 0.35	35%
19.	$\frac{3}{10}$	**a.** 0.3	**b.** 30%
20.	**a.** $\frac{2}{25}$	0.08	**b.** 8%
21.	**a.** $\frac{2}{5}$	**b.** 0.4	40%
22.	$\frac{3}{25}$	**a.** 0.12	**b.** 12%
23.	**a.** $\frac{11}{100}$	0.11	**b.** 11%
24.	**a.** $\frac{3}{50}$	**b.** 0.06	6%
25.	$\frac{99}{100}$	**a.** 0.99	**b.** 99%

Set JJ

Write each ratio as a reduced fraction.

1. 4 to 6 $\frac{2}{3}$ **2.** 6 to 4 $\frac{3}{2}$ **3.** 3 to 5 $\frac{3}{5}$ **4.** 5 to 10 $\frac{1}{2}$

5. 10 to 6 $\frac{5}{3}$ **6.** 8 to 10 $\frac{4}{5}$ **7.** 8 to 12 $\frac{2}{3}$ **8.** 12 to 9 $\frac{4}{3}$

9. 6 to 9 $\frac{2}{3}$ **10.** 9 to 6 $\frac{3}{2}$ **11.** 60 to 100 $\frac{3}{5}$ **12.** 90 to 100 $\frac{9}{10}$

There are 12 girls and 15 boys.

13. What is the boy-girl ratio? $\frac{5}{4}$

14. What is the girl-boy ratio? $\frac{4}{5}$

There are 200 dogs and 450 cats.
15. What is the dog-cat ratio? $\frac{4}{9}$
16. What is the cat-dog ratio? $\frac{9}{4}$

There are 600 students and 24 teachers.
17. What is the student-teacher ratio? $\frac{25}{1}$
18. What is the teacher-student ratio? $\frac{1}{25}$

There are 4 red marbles, 6 white marbles, and 10 blue marbles.
19. What is the red-white ratio? $\frac{2}{3}$
20. What is the red-blue ratio? $\frac{2}{5}$
21. What is the white-red ratio? $\frac{3}{2}$
22. What is the white-blue ratio? $\frac{3}{5}$
23. What is the blue-red ratio? $\frac{5}{2}$
24. What is the blue-white ratio? $\frac{5}{3}$
25. What is the white–not white ratio? $\frac{3}{7}$

Set KK

Complete each proportion.

1. $\dfrac{3}{4} = \dfrac{a}{20}$ 15 **2.** $\dfrac{4}{6} = \dfrac{12}{b}$ 18 **3.** $\dfrac{c}{6} = \dfrac{8}{12}$ 4 **4.** $\dfrac{4}{d} = \dfrac{24}{60}$ 10

5. $\dfrac{8}{10} = \dfrac{12}{e}$ 15 **6.** $\dfrac{9}{10} = \dfrac{f}{1000}$ 900 **7.** $\dfrac{g}{4} = \dfrac{27}{36}$ 3 **8.** $\dfrac{15}{h} = \dfrac{30}{42}$ 21

9. $\dfrac{12}{15} = \dfrac{i}{25}$ 20 **10.** $\dfrac{7}{8} = \dfrac{700}{j}$ 800 **11.** $\dfrac{k}{60} = \dfrac{4}{5}$ 48 **12.** $\dfrac{12}{l} = \dfrac{60}{100}$ 20

13. $\dfrac{3}{25} = \dfrac{m}{100}$ 12 **14.** $\dfrac{3}{8} = \dfrac{9}{n}$ 24 **15.** $\dfrac{p}{20} = \dfrac{35}{100}$ 7 **16.** $\dfrac{12}{q} = \dfrac{10}{20}$ 24

17. $\dfrac{27}{50} = \dfrac{r}{100}$ 54 **18.** $\dfrac{6}{18} = \dfrac{7}{s}$ 21 **19.** $\dfrac{t}{4} = \dfrac{75}{100}$ 3 **20.** $\dfrac{27}{u} = \dfrac{15}{20}$ 36

Set LL

Name each polygon.

1.

pentagon

2.

octagon

3.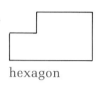

hexagon

Name each quadrilateral.

4.

trapezoid

5.

parallelogram

6. A parallelogram with all sides equal in length   rhombus

7. A rectangular parallelogram with all sides equal in length
square

Name each triangle by sides.

8.

isosceles

9.

equilateral

10.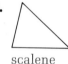

scalene

Name each triangle by angles.

11.

acute

12.

right

13.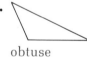

obtuse

Name each solid.

14.

cylinder

15.

cube

16.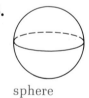

sphere

For each shape give (a) the perimeter and (b) the area. (All units are cm.)

17.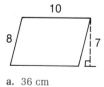

a. 36 cm
b. 70 sq. cm

18.

a. 54 cm
b. 126 sq. cm

19.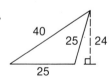

a. 90 cm
b. 300 sq. cm

20.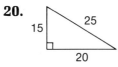
a. 60 cm
b. 150 sq. cm

21.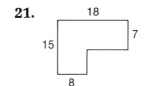
a. 66 cm
b. 190 sq. cm

22.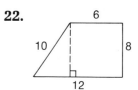
a. 36 cm
b. 72 sq. cm

23.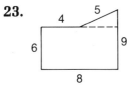
a. 32 cm
b. 54 sq. cm

24.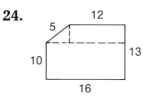
a. 56 cm
b. 202 sq. cm

25.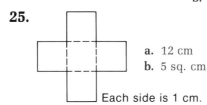
a. 12 cm
b. 5 sq. cm

Each side is 1 cm.

Set MM

Write each quotient in three forms.

PROBLEM	WITH REMAINDER	FRACTION	DECIMAL
1. $100 \div 8 =$	12 r4	$12\frac{1}{2}$	12.5
2. $50 \div 4 =$	12 r2	$12\frac{1}{2}$	12.5
3. $56 \div 10 =$	5 r6	$5\frac{3}{5}$	5.6
4. $65 \div 10 =$	6 r5	$6\frac{1}{2}$	6.5
5. $63 \div 12 =$	5 r3	$5\frac{1}{4}$	5.25
6. $72 \div 5 =$	14 r2	$14\frac{2}{5}$	14.4
7. $49 \div 4 =$	12 r1	$12\frac{1}{4}$	12.25
8. $38 \div 8 =$	4 r6	$4\frac{3}{4}$	4.75
9. $47 \div 4 =$	11 r3	$11\frac{3}{4}$	11.75
10. $146 \div 8 =$	18 r2	$18\frac{1}{4}$	18.25

PROBLEM	WITH REMAINDER	FRACTION	DECIMAL
11. $390 \div 20 =$	19 r10	$19\frac{1}{2}$	19.5
12. $625 \div 10 =$	62 r5	$62\frac{1}{2}$	62.5
13. $432 \div 5 =$	86 r2	$86\frac{2}{5}$	86.4
14. $650 \div 8 =$	81 r2	$81\frac{1}{4}$	81.25
15. $325 \div 20 =$	16 r5	$16\frac{1}{4}$	16.25
16. $562 \div 8 =$	70 r2	$70\frac{1}{4}$	70.25
17. $530 \div 40 =$	13 r10	$13\frac{1}{4}$	13.25
18. $375 \div 50 =$	7 r25	$7\frac{1}{2}$	7.5
19. $240 \div 100 =$	2 r40	$2\frac{2}{5}$	2.4
20. $534 \div 10 =$	53 r4	$53\frac{2}{5}$	53.4
21. $123 \div 4 =$	30 r3	$30\frac{3}{4}$	30.75
22. $321 \div 8 =$	40 r1	$40\frac{1}{8}$	40.125
23. $123 \div 6 =$	20 r3	$20\frac{1}{2}$	20.5
24. $321 \div 12 =$	26 r9	$26\frac{3}{4}$	26.75
25. $987 \div 24 =$	41 r3	$41\frac{1}{8}$	41.125

Set NN

1. What is 25% of 100? 25   **2.** What is 25% of 200? 50
3. What is 25% of 400? 100   **4.** What is 25% of 40? 10
5. What is 25% of 20? 5   **6.** What is 20% of 100? 20
7. What is 20% of 50? 10   **8.** What is 20% of 5? 1
9. What is 50% of 200? 100   **10.** What is 50% of 100? 50
11. What is 50% of 50? 25   **12.** What is 50% of 12? 6
13. What is 60% of 100? 60   **14.** What is 60% of 200? 120
15. What is 60% of 50? 30   **16.** What is 60% of 25? 15
17. What is 75% of 100? 75   **18.** What is 75% of 400? 300
19. What is 75% of 40? 30   **20.** What is 75% of 4? 3
21. What is 100% of 100? 100 **22.** What is 100% of 50? 50

23. What is 10% of 50?   5   **24.** What is 10% of 100?   10
25. What is 10% of 60?   6

Set OO

Write the decimal numeral named.
  1. $(6 \times 10^3) + (2 \times 10^2)$ 6200
  2. (5×10^4) 50,000
  3. $(3 \times 10^2) + (2 \times 10^1)$ 320
  4. $(5 \times 10^2) + (9 \times 10^0)$ 509
  5. $(6 \times 10^4) + (3 \times 10^2) + (4 \times 10^1)$ 60,340
  6. (7×10^1) 70
  7. $(6 \times 10^1) + (4 \times 10^0)$ 64
  8. $(5 \times 10^5) + (3 \times 10^4)$ 530,000
  9. $(4 \times 10^3) + (8 \times 10^2) + (2 \times 10^1)$ 4820
10. $(6 \times 10^4) + (2 \times 10^3)$ 62,000
11. (8×10^0) 8
12. $(7 \times 10^2) + (6 \times 10^1) + (5 \times 10^0)$ 765
13. $(9 \times 10^4) + (8 \times 10^1)$ 90,080

Write in expanded notation with exponents.

14. 500	**15.** 4600	**16.** 60	**17.** 750
(5×10^2)	$(4 \times 10^3) + (6 \times 10^2)$	(6×10^1)	$(7 \times 10^2) + (5 \times 10^1)$
18. 90,000	**19.** 10,500	**20.** 9	**21.** 701
(9×10^4)	$(1 \times 10^4) + (5 \times 10^2)$	(9×10^0)	$(7 \times 10^2) + (1 \times 10^0)$
22. 21	**23.** 45,000	**24.** 7010	**25.** 400,000
$(2 \times 10^1) + (1 \times 10^0)$	$(4 \times 10^4) + (5 \times 10^3)$	$(7 \times 10^3) + (1 \times 10^1)$	(4×10^5)

Set PP

  1. Six is what percent of 12?   50%
  2. Six is what percent of 10?   60%
  3. Six is what percent of 8?   75%
  4. Six is what percent of 6?   100%
  5. Twelve is what percent of 120?   10%
  6. Twelve is what percent of 60?   20%
  7. Twelve is what percent of 48?   25%

8. Twelve is what percent of 24?　50%
9. Twelve is what percent of 12?　100%
10. Twelve is what percent of 16?　75%
11. Twelve is what percent of 15?　80%
12. Twelve is what percent of 100?　12%
13. What percent of 100 is 8?　8%
14. What percent of 10 is 8?　80%
15. What percent of 16 is 8?　50%
16. What percent of 8 is 8?　100%
17. What percent of 80 is 8?　10%
18. What percent of 100 is 20?　20%
19. What percent of 80 is 20?　25%
20. What percent of 40 is 20?　50%
21. What percent of 25 is 20?　80%
22. What percent of 20 is 20?　100%
23. What percent of 15 is 3?　20%
24. What percent of 100 is 30?　30%
25. What percent of 40 is 30?　75%

Set QQ

1. $-3 + (-4)$　-7　　**2.** $(-5) + (-8)$　-13　**3.** $+3 + (-5)$　-2
4. $3 + (-8)$　-5　　**5.** $-3 + 4$　$+1$　　**6.** $(-4) + (+3)$　-1
7. $-7 + 6$　-1　　**8.** $(-6) + (+6)$　0　　**9.** $5 + (-11)$　-6
10. $(+5) + (+7)$　$+12$　**11.** $-12 + (-12)$　-24　**12.** $-12 + 12$　0
13. $(-12) + (+15)$　$+3$　**14.** $(+8) + (-1)$　$+7$　**15.** $(-8) + (+1)$　-7
16. $-5 + 8$　$+3$　**17.** $+15 + (-18)$　-3　**18.** $-25 + (-30)$　-55
19. $(+15) + (-20)$　-5　**20.** $8 + (-16)$　-8　**21.** $-9 + 15$　$+6$
22. $(-6) + (+20)$　$+14$　**23.** $(-20) + (-12)$　-32　**24.** $(+6) + (-4)$　$+2$
25. $8 + (-18)$　-10

Set RR

1. $3 - (-5)$　$+8$　　**2.** $-3 - 5$　-8　　**3.** $-3 - (-5)$　$+2$
4. $(-5) - (-3)$　-2　**5.** $(-5) - (+3)$　-8　**6.** $5 - (-6)$　$+11$

7. $7 - (-12)$ +19 **8.** $-7 - (-12)$ +5 **9.** $(-7) - (+12)$
−19
10. $-12 - 7$ −19 **11.** $-12 - (-7)$ −5 **12.** $12 - (-7)$ +19
13. $(-12) - (+7)$ −19 **14.** $-6 - (-6)$ 0 **15.** $(+6) - (-6)$ +12
16. $(-10) - (+5)$ −15 **17.** $-10 - (-5)$ −5 **18.** $-5 - 10$ −15
19. $-5 - (-10)$ +5 **20.** $10 - (-5)$ +15 **21.** $-12 - (-8)$ −4
22. $(-12) - (+8)$ −20 **23.** $-8 - 12$ −20 **24.** $-8 - (-12)$ +4
25. $8 - (-12)$ +20

Set SS

1. Fifty is $\frac{1}{2}$ of what? 100 **2.** Forty is $\frac{1}{4}$ of what? 160
3. Thirty is $\frac{1}{5}$ of what? 150 **4.** Twenty is $\frac{1}{10}$ of what? 200
5. Ten is $\frac{2}{5}$ of what? 25 **6.** Twenty is $\frac{2}{3}$ of what? 30
7. Thirty is $\frac{3}{4}$ of what? 40 **8.** Forty is $\frac{2}{5}$ of what? 100
9. Fifty is $\frac{5}{8}$ of what? 80 **10.** Twelve is $\frac{1}{4}$ of what? 48
11. Twelve is $\frac{3}{4}$ of what? 16 **12.** Twelve is $\frac{1}{3}$ of what? 36
13. Twelve is $\frac{3}{5}$ of what? 20 **14.** Sixty is 0.1 of what? 600
15. Sixty is 0.2 of what? 300 **16.** Sixty is 0.3 of what? 200
17. Sixty is 0.4 of what? 150 **18.** Sixty is 0.5 of what? 120
19. Sixty is 0.6 of what? 100 **20.** Thirty-five is 0.7 of what? 50
21. Forty is 0.8 of what? 50 **22.** Forty-five is 0.9 of what? 50
23. Ten is 0.01 of what? 1000 **24.** Twenty is 0.25 of what? 80
25. Thirty is 0.75 of what? 40

Set TT

1. $\sqrt{36}$ 6 **2.** $\sqrt{4}$ 2 **3.** $\sqrt{81}$ 9 **4.** $\sqrt{49}$ 7
5. $\sqrt{121}$ 11 **6.** $\sqrt{16}$ 4 **7.** $\sqrt{64}$ 8 **8.** $\sqrt{100}$ 10
9. $\sqrt{25}$ 5 **10.** $\sqrt{9}$ 3 **11.** $\sqrt{144}$ 12 **12.** $\sqrt{1}$ 1

Each square root is between which two consecutive whole numbers?
13. $\sqrt{10}$ 3, 4 **14.** $\sqrt{30}$ 5, 6 **15.** $\sqrt{40}$ 6, 7 **16.** $\sqrt{50}$ 7, 8 **17.** $\sqrt{60}$ 7, 8

18. $\sqrt{70}$ 8, 9 **19.** $\sqrt{80}$ 8, 9 **20.** $\sqrt{90}$ 9, 10 **21.** $\sqrt{5}$ 2, 3 **22.** $\sqrt{2}$ 1, 2
23. $\sqrt{75}$ 8, 9 **24.** $\sqrt{12}$ 3, 4 **25.** $\sqrt{24}$ 4, 5

Set UU

Complete the chart for each circle. (Use $\pi = 3.14$.)

RADIUS	DIAMETER	CIRCUMFERENCE	AREA
1 in.	**1.** 2 in.	**2.** 6.28 in.	**3.** 3.14 sq. in.
4. 2 ft.	4 ft.	**5.** 12.56 ft.	**6.** 12.56 sq. ft.
3 cm	**7.** 6 cm	**8.** 18.84 cm	**9.** 28.26 sq. cm
10. 4 m	8 m	**11.** 25.12 m	**12.** 50.24 sq. m
5 mm	**13.** 10 mm	**14.** 31.4 mm	**15.** 78.5 sq. mm
16. 6 yd.	12 yd.	**17.** 37.68 yd.	**18.** 113.04 sq. yd.
10 km	**19.** 20 km	**20.** 62.8 km	**21.** 314 sq. km
22. 50 mi.	100 mi.	**23.** 314 mi.	**24.** 7850 sq. mi.

25. What special word is used to describe the perimeter of a circle?   circumference

Set VV

1. Ten is 10% of what?   100
2. Ten is 20% of what?   50
3. Ten is 40% of what?   25
4. Ten is 25% of what?   40
5. Ten is 5% of what?   200
6. Ten is 2% of what?   500
7. One hundred is 100% of what?   100
8. Twenty is 20% of what?   100

9. Twenty is 10% of what?   200
10. Twenty is 1% of what?   2000
11. Six is 10% of what?   60
12. Six is 25% of what?   24
13. Six is 1% of what?   600
14. Five is 100% of what?   5
15. Fifty is 50% of what?   100
16. Fifty is 100% of what?   50
17. Fifty is 10% of what?   500
18. Six is 75% of what?   8
19. Ten is 100% of what?   10
20. Fifty is 25% of what?   200
21. Forty is 100% of what?   40
22. Forty is 40% of what?   100
23. Forty is 50% of what?   80
24. Forty is 25% of what?   160
25. Twenty is 25% of what?   80

Set WW

1. $(-3)(+4)$ -12
2. $(-8)(-12)$ $+96$
3. $(+12)(-15)$ -180
4. $(-18)(-20)$ $+360$
5. $(+21)(-7)$ -147
6. $(-8)(+17)$ -136
7. $(-25)(-15)$ $+375$
8. $(+18)(+5)$ $+90$
9. $(-7)(+43)$ -301
10. $(-12)(-24)$ $+288$
11. $(+6)(-18)$ -108
12. $(-9)(-25)$ $+225$
13. $(-24) \div (+2)$ -12
14. $(-24) \div (-3)$ $+8$
15. $-72 \div 8$ -9
16. $400 \div (-5)$ -80
17. $(-234) \div (-6)$ $+39$
18. $(-144) \div (+12)$ -12
19. $-125 \div 25$ -5
20. $(-375) \div (-15)$ $+25$
21. $\dfrac{-144}{8}$ -18
22. $\dfrac{400}{-20}$ -20
23. $\dfrac{-456}{-12}$ $+38$
24. $\dfrac{575}{-25}$ -23
25. $\dfrac{-360}{-120}$ $+3$

Set XX

Write the Arabic numeral.

1. XIX 19
2. XVIII 18
3. CCCXCI 391
4. CDXXIV 424
5. MCC 1200
6. LIV 54
7. CCLVI 256
8. MMCLX 2160
9. XCII 92
10. CDXXV 425
11. XXIV 24
12. CCXLIV 244
13. MCMLXIX 1969
14. MLXVI 1066
15. MCMLXXXIV 1984

Write the Roman numeral.

16. 63 LXIII
17. 89 LXXXIX
18. 101 CI
19. 651 DCLI
20. 44 XLIV
21. 99 XCIX
22. 156 CLVI
23. 490 CDXC
24. 2310 MMCCCX
25. 1907 MCMVII

Set YY

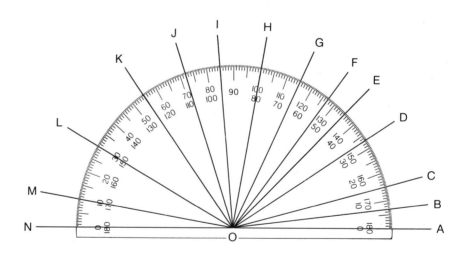

Using the protractor, find the measure of each of the following angles.

1. ∠AOB 7°
2. ∠AOE 45°
3. ∠AOH 80°
4. ∠AOK 125°
5. ∠AON 180°
6. ∠AOD 33°
7. ∠AOG 65°
8. ∠AOJ 108°
9. ∠AOM 170°
10. ∠AOC 15°
11. ∠AOF 53°
12. ∠AOI 95°
13. ∠AOL 150°
14. ∠NOL 30°
15. ∠NOI 85°
16. ∠NOF 127°
17. ∠NOC 165°
18. ∠NOK 55°
19. ∠NOH 100°
20. ∠NOE 135°
21. ∠NOM 10°
22. ∠NOJ 72°
23. ∠NOG 115°
24. ∠NOD 147°
25. ∠NOB 173°

Set ZZ

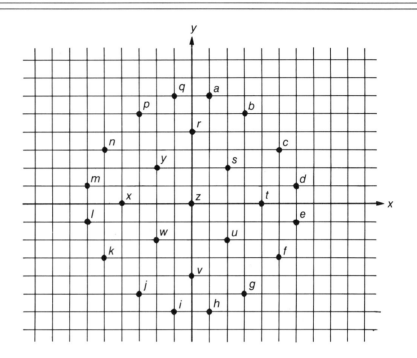

1. What is the name of the vertical axis? y-axis
2. What is the name of the horizontal axis? x-axis
3. What is the name given to the point at which the x and y axes cross? origin
4. What name do we use to describe the numbers that give the location of a point? coordinates

Which points above have these coordinates?

5. (3, 5) b   6. (0, 4) r   7. (−2, −2) w 8. (−4, 0) x
9. (2, −2) u 10. (5, −3) f 11. (−1, 6) q 12. (−3, −5) j

What are the coordinates of these points?

13. *l* (−6, −1)
14. *a* (1, 6)
15. *c* (5, 3)
16. *d* (6, 1)
17. *g* (3, −5)
18. *i* (−1, −6)
19. *k* (−5, −3)
20. *v* (0, −4)
21. *m* (−6, 1)
22. *p* (−3, 5)
23. *y* (−2, 2)
24. *s* (2, 2)
25. *z* (0, 0)

Calculator Exercise 1
(Following Lesson 15)

Write what is displayed after the equals sign is pressed.

1. (a) $8 - 0 = 8$ (b) $0 - 8 = -8$

2. (a) $6 - 5 = 1$ (b) $5 - 6 = -1$

3. (a) $15 - 7 = 8$ (b) $7 - 15 = -8$

4. (a) $20 - 8 = 12$ (b) $8 - 20 = -12$

5. (a) $46 - 25 = 21$ (b) $25 - 46 = -21$

6. (a) $32 - 8 = 24$ (b) $8 - 32 = -24$

7. (a) $121 - 11 = 110$ (b) $11 - 121 = -110$

8. (a) $537 - 164 = 373$ (b) $164 - 537 = -373$

9. (a) $1001 - 101 = 900$ (b) $101 - 1001 = -900$

10. (a) $37,286 - 8,934 = 28,352$ (b) $8,934 - 37,286 = -28,352$

Calculator Exercise 2
(Following Lesson 32)

Key in the number one million: 1,000,000. Then repeatedly divide by 10 and record the display each time the equals sign is pressed.

1. $1,000,000 \div 10 = 100,000$ 2. $\div 10 = 10,000$

3. $\div 10 = 1000$ 4. $\div 10 = 100$

5. $\div 10 = 10$ 6. $\div 10 = 1$

7. $\div\, 10 = 0.1$ **8.** $\div\, 10 = 0.01$

9. $\div\, 10 = 0.001$ **10.** $\div\, 10 = 0.0001$

Calculator Exercise 3 (Following Lesson 34)

Since $\frac{3}{10}$ means $3 \div 10$, for $\frac{3}{10}$ we key in

$$\boxed{3}\ \boxed{\div}\ \boxed{10}\ \boxed{=}$$

Write what is displayed after the equals sign is pressed.

1. $\dfrac{3}{10} = 0.3$ **2.** $\dfrac{7}{10} = 0.7$ **3.** $\dfrac{23}{100} = 0.23$

4. $\dfrac{9}{100} = 0.09$ **5.** $\dfrac{1}{100} = 0.01$ **6.** $\dfrac{321}{1000} = 0.321$

7. $\dfrac{21}{1000} = 0.021$ **8.** $\dfrac{1}{1000} = 0.001$ **9.** $\dfrac{10}{1000} = 0.01$

10. $\dfrac{100}{1000} = 0.1$

Calculator Exercise 4 (Following Lesson 46)

The calculator will simplify decimal numbers. Enter the digits given and write what is displayed when the equals sign is pressed.

1. $0000 = 0$ **2.** $0.0000 = 0$ **3.** $0001 = 1$

4. $1.0000 = 1$ **5.** $0.1000 = 0.1$ **6.** $01.100 = 1.1$

7. $01.0100 = {\scriptstyle 1.01}$ **8.** $100.001 = {\scriptstyle 100.001}$ **9.** $0.0001 = {\scriptstyle 0.0001}$

10. $.1 = {\scriptstyle 0.1}$

Calculator Exercise 5
(Following Lesson 51)

Write what is displayed after the equals sign is pressed. (Can you beat the calculator to the answer?)

1. $1.2 \times 10 = {\scriptstyle 12}$ **2.** $1.2 \times 100 = {\scriptstyle 120}$

3. $1.2 \times 1000 = {\scriptstyle 1200}$ **4.** $1.2 \div 10 = {\scriptstyle 0.12}$

5. $1.2 \div 100 = {\scriptstyle 0.012}$ **6.** $1.2 \div 1000 = {\scriptstyle 0.0012}$

7. $.012 \times 10 = {\scriptstyle 0.12}$ **8.** $.012 \times 100 = {\scriptstyle 1.2}$

9. $.012 \times 1000 = {\scriptstyle 12}$ **10.** $12 \div 10 = {\scriptstyle 1.2}$

11. $12 \div 100 = {\scriptstyle 0.12}$ **12.** $12 \div 1000 = {\scriptstyle 0.012}$

Index